W0107261

93 Advances in Polymer Science

Polymer Processing

Editor: M. L. Fridman

With contributions by
E. Yu. Bormashenko, M. L. Fridman, V. S. Levin,
S. L. Peshkovsky, A. Z. Petrosyan, V. D. Sevruk,
V. I. Tunkel

With 80 Figures and 13 Tables

Springer-Verlag Berlin Heidelberg GmbH

ISBN 978-3-662-15041-2   ISBN 978-3-540-46202-6 (eBook)
DOI 10.1007/978-3-540-46202-6

Library of Congress Catalog Card Number 61-642

© Springer-Verlag Berlin Heidelberg 1990
Originally published by Springer-Verlag Berlin Heidelberg New York in 1990
Softcover reprint of the hardcover 1st edition 1990

2152/3020-543210 — Printed on acid-free paper

Editors

Editor's Foreword

Processing of polymers is one of the most important and rapidly progressing industries the theoretical basis of which is gaining respect. Plastic processing technology and the development of the relevant equipment can not be considered today merely as a set of empirical approaches employed to manufacture high-quality products. Such approaches would mean a "retreat" and a reduction of activities in this field to mediocrity and enclose the process within the framework of just an accumulation of skills and knowledge about "standard" cases of manufacturing experience.

The fact is that, in the long history of polymer processing, engineering and design has always been ahead of theory. The development of screwless (or disc-type) extruders, innovative for their time, on the basis of the earlier-discovered normal stress effect (which received the name of the Weissenberg effect) was, apparently, one of the few exclusions. However, this example has clearly demonstrated the potential and the role of theoretical research in the progress of technology.

The situation in the field of processing has been changing during recent years. A number of research projects have contributed to the development of theoretical premises not only for the intensification and optimization of the known processes but also for development of new technologies and products. This can be easily exemplified: successful studies of the orientation crystallization of polymers have provided a basis for the commercial production of superstrong materials (fibers, films, etc.); data on changes in the structure and mechanical properties of polymers under extrusion have been used to develop the technology and equipment for manufacturing of film threads and fibers (including the fibrillated ones); development of the composite shear theory and demonstration of the possibility of reducing the viscosity of melts and intensifying their flow when the polymer is in the combined-stress state occurring at low- and high-frequency oscillations (vibrations) of the molding tool have stimulated the search for designs of machinery with an increased molding rate attained on the basis of the above mentioned effect; development of mathematical models of many operations and processes, generally, has become an inallienable part of the automatization and computerization of the modern production lines; finally, the theory has provided a reliable basis for computer-aided engineering of the main working elements of extruders, casting machines, extrusion heads, and molding tools.

The list of similar achievements could easily be extended but even the above cited examples are sufficient to conclude that the further progress of polymer processing depends strongly on the successful development and improvement of the theoretical basis of technology. We hope that this publication will make a practical contribution to the elimination of the still-existing gap between "science" and "technology".

Certainly, it is impossible to review all the new developments in polymer processing in one book even if we limit ourselves to the short period of the last 5–10 years. This, however, had been planned neither by the publisher nor by by the editor. We have collected under one cover four reviews every one of which, we believe, illustrates different routes of the progress of scientific knowledge in the sphere of polymer processing and demonstrates various stages in the development of this knowledge and its practical application.

Thus the review on the problem of molten polymer extension (M. L. Fridman & V. D. Sevruk) describes not only achievements of the theory and experiments but also new approaches to estimation of the quality of thermoplastics on the basis of melt extension tests under constant-force conditions. During the next few years we may expect development of new instrumentation and standard procedures for testing raw materials used to manufacture films, fibres, etc. In other words, these data are available for practical application and it is high time we took advantage of them.

Another review on the problem of polymer molding under vibration effects is, in our view, fundamentally important. The idea of this publication is not only to analyze and generalize results of combined-shear studies but also to attract attention of experts to the two latest achievements: theoretical and experimental corroboration of the efficiency of physical effects upon molten polymers in molding processes, and discovery of a new phenomenon — acoustic cavitation of molten polymers — which had not been predicted by the cavitation theory for fluids with a viscosity that high. Practical applications of this effect in manufacturing technologies have still to be developed. However it is clear right now that it has a high potential, since the melt cavitation conditions permit us to adjust the melt's rheological properties, to attain "dosed" mechanical destruction, excite active radicals in the polymer, disperse fillers, make new mixtures and alloys of polymers. We believe that the effect of acoustic cavitation opens a possibility of creating a new branch of plastic processing. However its present status leaves room for improvement: the effect has been discovered and extensive scientific information is available, but it lacks the technology, or a number of technologies, which could be developed on that basis. We expect that these technologies will be characterized by low power consumption and high efficiency.

The situation is different in the case of processing of polymer materials at low pressure. Technology has been rapidly advancing in the sphere in recent years: centrifugal and rotary molding methods are widely used, the RIM-technology is generally accepted, polymer paste casting is used, etc. As regards the theory of the low pressure molding of polymer materials, it has not been completed so far and here we are facing again the situation when technology has leaped ahead of theory. We wished to demonstrate (M. L. Fridman, A. Z. Petrosyan, V. S. Levin, E. Yu. Bormashenko), using the example of a systematic review of the data on analysis and mathematical modelling of the casting of polymer pastes or plastisols, that science can provide a basis for automatization and optimization of the process, offer procedures for the analysis of optimum geometrical parameters of molding tools and temperature conditions of molding to be used by engineers and designers. An important point is that the developed approaches can be applied later to the analysis of the molding of low-viscous thermoplastics and other low-pressure processing technologies.

Finally, we suppose that the review on the aminoplastic granulation technology (V. I. Tunkel, M. L. Fridman) illustrates another level of the development of processing. The analysis makes it clear that the situation in this case is quite the opposite: a scientific basis of the granulation process is available but the optimum engineering solutions are still to be found. Therefore we have focused on the comparative analysis of machinery and equipment for the process. It should be noted that although aminoplastic nodulizing technology has been practically abandoned during recent years, the interest for this technology can be reanimated, apparently, due to the new engineering approaches.

Certainly, the above mentioned does not give a full answer to the question, why these four articles have been included into this collection. It is natural that they have been selected primarily with regard to the scientific and engineering interests of the editor (who is also one of the authors) and include works carried out during recent years together with junior colleagues.

In conclusion I wish to express my gratitude to Springer Verlag and, personally, to the Chemistry Editor Dr. R. Stumpe for granting us the opportunity to use a special issue of this much respected journal to disseminate our knowledge.

Moscow, October 1989   Professor M. Fridman

Table of Contents

Extension of Molten Polymers

M. L. Fridman, V. D. Sevruk

USSR Research Institute of Plastic Materials

Perovsky pr. 35, Moskow 111112, USSR

This contribution reviews the major results of studies of the extension of molten polymers which have been carried out recently. The authors discuss systematically basic regularities of the extension of molten polydisperse polymers including the uniform extension and its development in time (at a constant strain velocity and at constant extrusion force). The article also considers the dependency of stress and strain velocity upon elastic strain, stress and strain relaxation processes; the major differences in the variations of effective viscosity under extension are pointed out. The authors describe the effect of polymer fluid flow retardation under high elastic strains.

The article reviews the latest achievements in the sphere of theoretical descriptions of the molten polydisperse polymers and gives various molecular-kinetic models of extension.

Also described are some important technological applications in the processing of polymers, including the methods of examination and verification of the properties of raw materials by means of tests in which molten polymers are extended at a constant force, and molding of sleeve-type and flat films.

The analysis has corroborated that the extension experiments were highly informative and important for science and technology.

Advances in Polymer Science 93
© Springer-Verlag Berlin Heidelberg 1990

1 Introduction

Practically all technologies of polymer processing are associated with extension of melts. In many cases it is an important but still "associate" effect as, for example, extension of melts in the zone close to the inlet to the molding part of extrusion heads, extension of jets during filling of molds in the course of pressurized casting. However, in a number of the major modern technologies of polymer processing the extension of melts is not an "effect" but the primary manufacturing operation principally critical for the process on the whole or, at least, for the production of special-purpose semifinished products and products with a present range of properties and geometrical dimensions. This group of processes includes, for example, molding of fibres, extrusion-inflated (sleeve) films with preset width and thickness, manufacturing of flat film blanks for production of thin and ultrathin uniaxially and biaxially oriented films, uniaxially oriented, flat threads ("refining") and fibrillated (splitted) elementary threads for further processing into twisted (braids, binder twines, ropes, etc.) and wattled items. The development of scientific fundamentals and improvement of the above-mentioned technologies requires a detailed knowledge and understanding of the behavior of molten polymers not only under shear strain (flow) but under tension, first of all at the outlet of the molding tools, which is usually called "spinneret drawing". It should be emphasized that the spinneret drawing in industrial technologies is complicated by a number of nontrivial rheological and physical/chemical phenomena, such as jet swelling at the outlet from the channel ("Barrus-effect"), nonisothermicity of flow and extension, development of chemical reactions under stress (thermomechanical destruction) [1]. We must admit that an understanding of these processes which could provide a basis for accurate and comprehensive calculations of the operation of molten polymer extension from molding tools has not been reached so far.

However, the applied importance of research into extension of melts is far from being limited by the above-mentioned aspects. Thus, for example, technological properties of polymers, mixtures, and compositions based thereon have been so far classified and compared primarily in terms of rheological (first of all viscous) properties of melts under shear strain conditions which is not only "poor" from the point of view of information but is absolutely noncharacteristic of cases when, as mentioned above, the principally important technological stage is not the shear flow but extension (jet drawing). Development of another approach to analysis of "adaptibility" and comparison of the properties of polymeric raw materials of different types (and even of different batches of polymers of one and the same brand) on the basis of extension tests of melts in recent years has become a problem of current concern for rheology, polymer-material study, and processing technology [2].

At first we have deliberately focused on the applied (technological) importance of the study of melt behavior under extension since the theoretical importance of the analysis of melt extension for polymer physics and mechanics can be regarded as already generally recognized. The scientific "success" and recognition of melt extension stems, we believe, from several fundamental causes, major of which are as follows. The geometrical pattern of deformation (shear, twisting, tension, etc.) is not very important for mechanics of the usual solid bodies since there is a well-known and multiply verified connection (linear Hooke's mechanics) between the main (if not been all) characteristics. It also holds true, on the whole, in case of a purely viscous fluid: shear flows in such media (Newton's law) are of major interest from the practical point of view, and viscosity under tension equals triple shear viscosity (Trutone's principle) and these laws characterize sufficiently the behavior of molten polymers in the linear strain area [3,4]. The situation is different in the area of their nonlinear behavior and the available characteristics of strain under shear do not necessarily permit to prognosticate the behavior of a material under extension. Therefore the study of regularities in the behavior of polymer fluids under extension is of general fundamental importance for construction of rheological models which describe and explain the behavior of these materials.

Another principal problem, beside the accumulation of significant reversible strains, which has stipulated theoretical and experimental interest in the discussed area of physics and mechanics of polymer fluids is the establishment of extension limits, i.e., to which extent polymer jets can be extended (ultimate, or rupture, extension rate). In reality the absolute limit (ultimate degree of extension of a macromolecular tangle when it turns into a stretched chain) is not reached, a jet breaks much earlier due to interslipping of macromolecules and the moment of rupture is determined by the strain velocity. A study of regularities in the accumulation of significant reversible strains has revealed (see, for example, Ref. [5]) that there is a dependency of ultimate strains ε_e^* upon longitudinal strain velocity $\dot{\varepsilon}$ given schematically in Fig. 1. At $\dot{\varepsilon} \sim \dot{\varepsilon}_f$ elastic strains, generally speaking, are low but complete critical strains (continuous curve in Fig. 1) are unlimitedly large since at $\dot{\varepsilon} < \dot{\varepsilon}_f$ plastic strains grow without increase in the elastic component (it relaxes completely and the polymer behaves as a usual viscous fluid). In this case the jet, certainly, is also destroyed but the mechanism of destruction is not the one specific for polymers. The situation alters after $\dot{\varepsilon}_f$. At $\dot{\varepsilon} > \dot{\varepsilon}_f$, with increasing strain velocity (and, respectively, that of stresses), elastic strains grow and viscous flow strains drop. At $\dot{\varepsilon} > \dot{\varepsilon}_f$ the extension obligatorily results in a break (this has been proven clearly and comparatively long ago, for example, see Ref. [6]), i.e., a steady flow can not be attained; point b in the diagram given in Fig. 1 corresponds to an approximate equality of reversible and irreversible strains; in the area of $\dot{\varepsilon}_f < \dot{\varepsilon} < \dot{\varepsilon}_f$ the material being in the flow state is extended but after point b the elastic strains become increasingly dominating (although the flow is still noticeable) until, finally, at $\dot{\varepsilon} > \dot{\varepsilon}_e$ the flow can be practically neglected (polymer jet behaves like rubber), the extension is carried out under conditions of a forced high-elastic state and ε_e^* corresponds to rupture strain, the theoretical maximum of which is reached at point d at a certain velocity $\dot{\varepsilon}_g$ sufficient for total

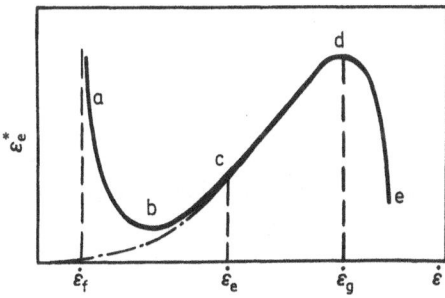

Fig. 1. Ultimate strains $\dot{\varepsilon}_e^*$ versus gradient of longitudinal velocity $\dot{\varepsilon}$ in extension (explanations are given in the text)

straightening of the macromolecular tangle. The section d—e represent a strain of the polymer in which the higher the strain velocity, the lower relaxation effects and after point e they become practically completely suppressed just as if the polymer were glass.

Proceeding from the above-described pattern we can formulate two problems — different in physical sense and in equipment employed for implementation — in the manufacturing of oriented products [7]. The first one deals with attaining of maximum values of $\dot{\varepsilon}_e^*$; this corresponds to implementation of such velocities and temperatures at which $\dot{\varepsilon}$ approaches $\dot{\varepsilon}_g$. Permissible strains are always limited here, flow does not take place, and there is a danger of breaking. This problem corresponds to modern high-speed processes of molding of chemical fibers (see, e.g., a recent review [8] dealing with achievements in this field). The second problem is the attainment of very high extension degrees while maximum possible $\dot{\varepsilon}_g^*$ are reached. The limit of this corresponds to strain velocity $\dot{\varepsilon}_f$. A similar problem occurs in manufacturing of films by method of extrusion with sleeve inflation when it is required to extend (draw and inflate) the melt with the help of a circular spinneret so that the cross section of the blank was reduced by a factor of 20 to 50 (and even more) and to avoid the potential danger of sleeve breaking.

However this publication has been stipulated not only by the general theoretical and applied importance of the extension of polymer melts. A number of reviews dealing with these problems has been published recently (see, for example, Refs. [9,10]) These publications cover a sufficiently wide scope of literature. However, they fail, generally, to consider works on extension of polydisperse polymers carried out in the USSR as well as the achievements in the developing field of extension of monodisperse polymers. Note that mono- and polydisperse polymers differ significantly from one another in terms of rheological behavior, at least because of their different degrees of elastic strain: elastic strains in monodisperse polymers are very low, while elastic strains in polydisperse are gigantic.

We believe that it would be useful not only to generalize the latest achievements in the rheology of extension of mono- and polydisperse systems, but also to assume and analyze some specific technological applications. Earlier works, which have become almost "classical" are referred to only for chronological accuracy and completeness of the view.

2 Major Regularities in Extension of Molten Polymers

2.1 Some Definitions

Rheological experiments with molten polymers are normally arranged to investigate into the homogeneous inertialess extension under isothermal conditions.

Usually, investigations deal with uniaxial tension in a cylindrical sample which is easier to produce in experiments than other types of tension (see Fig. 2). In this case the strain velocity tensor is:

$$e = \varkappa \operatorname{diag} \{1, -{}^1/_2, -{}^1/_2\} \tag{1}$$

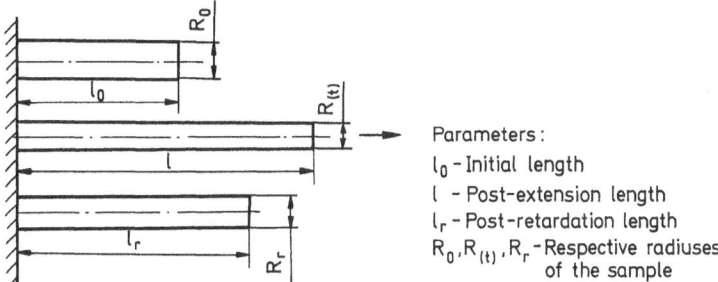

Parameters:
l_0 - Initial length
l - Post-extension length
l_r - Post-retardation length
$R_0, R_{(t)}, R_r$ - Respective radiuses
of the sample

Fig. 2. Diagram of uniform extension and subsequent retardation of cylindrical polymer sample

In the cylindrical system of coordinates v, z, φ (axis z is directed along the sample) components of velocity are expressed as:

$$v_z = \varkappa(t)z, \qquad v_r = -\frac{\varkappa(t)}{2} r, \qquad v_\varphi = 0 \tag{2}$$

If we neglect the effect of surface tension, there is only one component of stress tensor $\sigma = \sigma_{zz} = F/\pi R^2$ (F is extending force) independent of z in case of uniaxial tension of a cylinder.

Elongation is:

$$\varepsilon = 1/l_0 = R_0^2/R^2 \tag{3}$$

where l_0 and R_0 are sample length and radius, respectively, at the starting moment of strain; l and R are length and radius at an arbitrary moment of extension (see Fig. 2). The second equality in Eq. (3) follows from the condition of the incompressibility of the media. According to the known dependency of total strain ε upon time t, the strain velocity $\varkappa$ is determined from the following formula:

$$\varkappa = d(\ln \varepsilon)/dt \tag{4}$$

Longitudinal strain is taken as "elastic recoil":

$$\alpha = 1/l_r = R_r^2/R^2 \tag{5}$$

or logarithm of this value. In Eq. (5) and l_r and R_r are, respectively, the length and the radius to which, at $t \to \infty$ (in experiments, this time is practically finite), tends an extended sample with length 1 and radius R after momentary lifting of tensile stress (see Fig. 2). Contraction occurs due to elastic energy accumulated in extension of the polymer fluid.

The value of irreversible longitudinal strain in experimental measurements is taken [11, 12] as:

$$\ln \beta = \ln \varepsilon - \ln \alpha, \qquad \beta = l_p/l_0 \tag{6}$$

In this case the velocity of irreversible longitudinal strain can be determined in the following way:

$$e_p = d(\ln \beta) \, dt \tag{7}$$

Equation (7) was introduced in Refs. [13,14] where its reasonability was proved when the expression was checked for correctness in the area of minor elastic strains.

2.2 Homogeneous Extension of Polydisperse Polymers

2.2.1 General Aspects

The very first experiments in extension of polymers in the viscous flow state [11,12,15] demonstrated that $\sigma/x \neq$ const, as it is the case in Newtonian media. The relationship $\sigma/e_p = 3\eta$, true in the linear area under arbitrary conditions of tensile strain [13,14], does not hold true in the area of significant elastic deformations [11,12].

Later experimental research into extension was focused on detailed studies of the behavior of the medium's effective viscosity and the search for new effects under different conditions of extension.

Prior to a detailed discussion of the obtained experimental data we briefly list the major events — from our point of view — in chronological order.

1. *Ballman RL* [16]. Viscosity under stationary tension approaches the constant, while under shear at the same strain velocities it decreases rather intensively; unfortunately the experiment in extension was carried out within a limited range of strain velocities.

2. *Meissner Y* [17,18]. Construction of dependencies σ/x versus time t (here σ is tension stress, x is strain velocity registered in the course of extension). The relationship σ/x (effective viscosity) being independent of x in the linear region, such a processing of experimental data allowed to separate reliably the regions of nonlinear strain; unfortunately this work failed to attain the stationary flow area (a principal matter for understanding of the rheology of such media), apparently, because of the limited capacity of instrumentation.

3. *Radushkevitch BV, Fichman VD, Vinogradov GV* [19,20]. Stationary flow was attained under conditions of $x =$ const. Viscosity was increasing under stationary flow while the time during which the stationary flow was attained was reducing with increase in x; unfortunately the experiment was carried out rather roughly and the authors failed to obtain clear dependencies of σ/x upon t [17] therefrom.

4. *Munstedt H, Laun HM* [21]. Stationary flow was attained, which had not been revealed earlier in Refs. [17,18] (the experiments were carried out with the same polymer). The time during which the stationary flow was attained and viscosity depending upon x passed via a maximum. The possibility of viscosity drop has been noted earlier in a work by Cogswell [22]. The time during which the stationary flow was attained in the nonlinear area of flow [21] could exceed significantly the relaxation time determined in the linear regron.

5. *Prokunin AN* [23,24]. Flow deceleration effect. Two maxima in dependencies of force upon time undercoditions of $x =$ const and respective flow deceleration (which

was judged from the velocity of irreversible strain). The time during which the second maximum was attained could exceed significantly the characteristic relaxation time determined from shear experiments in the linear area.

6. *Prokunin AN, Sevruk VD* [25]. Two polymers characterized by not much different flow curves can differ qualitatively at homogeneous extension under conditions $\varkappa = \text{const}$ and $F = \text{const}$.

Let us discuss now in more detail the properties of polymers detected under homogeneous extension. We shall consider only two conditions: $\varkappa = \text{const}$ and $F = \text{const}$ (as the two most characteristic examples of flow with preset kinematics and stress).

2.2.2 Tensile Strain Development in Time. Extension at Constant Strain Velocity

This type of extension has been studied experimentally in many works [16–18,21,23, 26–33]; data from Refs. [18,19,24,29,31–33] will be discussed in detail in subsequent sections.

Dependencies of stress σ and elastic deformation α upon time t are measured in extension under conditions of $\varkappa = \text{const}$.

Let us consider the results of Refs. [23,26]. The experiments were carried out with polyisobutylene (PIB) II-20 at 22 °C with molecular weight (MW) $\sim 10^5$, maximum Newtonian viscosity $\eta_0 = 1.1 \times 10^6$ Pa · s and high-elasticity modulus $G_e = 1.57 \times 10^3$ Pa; relaxation time was $\theta = \eta/G \sim 10^3$ s. The constants were determined at shear.

Dependencies of dimensionless effective viscosity $\sigma/(\varkappa\eta)$ upon time t under extension with constant strain velocity $\varkappa$ for different $\varkappa$ are given in Fig. 3. At low $\varkappa$, when, at any t, the strain occurs in the linear region, the relationship $\sigma(t)/(\varkappa\eta)$ does not depend upon $\varkappa$ (see the lower line in Fig. 3) and equals 3 under stationary flow. This dependency coincides in the linear region of deformation with effective viscosity at shear $[3\sigma_{12}/(\dot\gamma\eta)]$. With increase in $\varkappa$, dependencies $\sigma/(\varkappa\eta)$ coincide with the linear strain curve only in the beginning (minor elastic strains) and depart from it strictly increasing with further increase of t. At high values of t the flow may reach stationary conditions ($\sigma/\varkappa = \text{const}$). The effective viscosity $\sigma/\varkappa$ and elastic strain α at stationary flow in the nonlinear strain regron grow with increase of $\varkappa$. Note also that the higher $\varkappa$, the earlier in the the stationary flow is attained. In the linear strain region the time during which the stationary flow is attained is of the same order of magnitude as value θ.

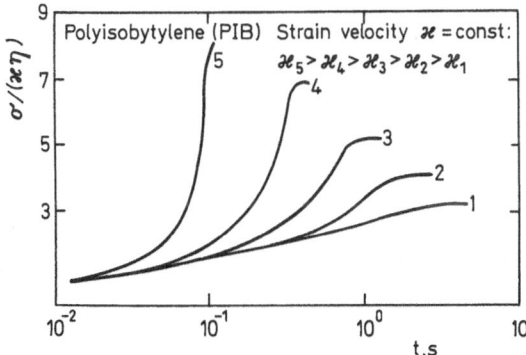

Fig. 3. Effective viscosity versus time at extension under conditions of constant strain velocities in molten polyisobutylene [23]

At high $\varkappa$, stationary flow was not attained. The stationary flow was not detected, perhaps, because of limitations imposed by the experimental procedure. This can be exemplified by Ref. [17] in which the stationary flow was not attained in the nonlinear region and which was detected only in Ref. [21] at high degrees of extension.

In works where the stationary flow was detected, the viscosity $\tilde{\lambda} = \sigma/\varkappa$ was, normally, either constant [16,20,31] or grew with increase of $\varkappa$ [18,19,24-27] above $3\eta_0$ (η_0 is the maximum shear viscosity) in the linear strain region. The intensive growth of viscosity $\tilde{\lambda}$ was detected, apparently, for the first time in Ref. [19]. The greatest increase in viscosity λ from 3η to 20η was observed in Refs. [18,21,30]. In the same works the viscosity $\tilde{\lambda}$ started to drop after the growth (this shall be discussed in detail below). Usually the relationship $\tilde{x}/\tilde{\eta}$ ($\tilde{\eta} = \sigma_{12}/\dot{\gamma}$ is effective viscosity at stationary shear) is taken at different strain velocities, which could vary by more than an order of magnitude (see, for example, Ref. [30]). The data on the existence of stationary flow and behavior of viscosity in this case, as well as descriptions of extended polymers are collected in Ref. [20].

Elastic strain in extension under conditions of $\varkappa = $ const in most works (see [18-20t, 24,26]) is a strictly increasing function of time with smoothly decreasing derivative $d(\ln \alpha)/dt$. In these cases the velocity of irreversible flow $e_p(t)$ strictly increases (exclusion detected in Refs. [23,24] is discussed below). At stationary flow the elastic strain is constant and the irreversible strain velocity is $e_p = \varkappa$. The higher $\varkappa$, the more the share of elastic component is at fixed general strain ε.

A specific feature of curves given in Fig. 3 is the fact that the time during which the stationary flow is attained strictly decreases with increase of $\varkappa$, while this is different in Refs. [18,24,34] (see below).

2.2.3 Extension at Constant Force

Extension at force $F = $ const corresponds to conditions with intensively increasing stress. The stationary flow, in this case, is obviously not attained, and deformation,

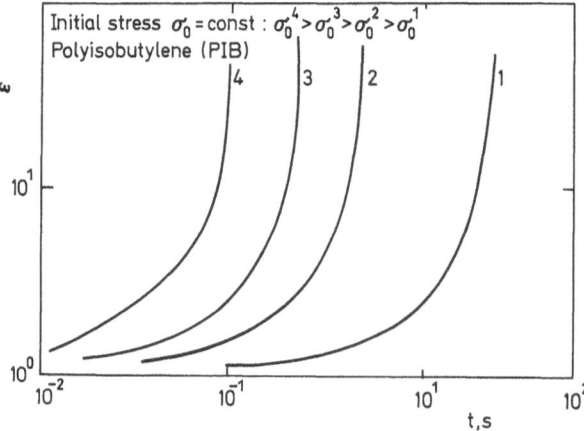

Fig. 4. Total strain ε versus time t in extension of polyisobutylene with different constant forces [23, 25]

sooner or later, leaves the linear region. The time of media staying in the linear region was estimated in Ref. [14].

In extension under conditions F = const, dependencies of total strain ε and elastic strain α upon time t are measured as usual. In Refs. [11-13, 15, 35-38] these experiments were arranged with polyisobutylene samples with different molecular weights and with low-density polyethylene (see below).

Dependencies of total strain $\varepsilon = l/l_0$ upon time t, in extension of polyisobutylene II-20 (USSR) at constant force F for different initial stresses $\sigma_0 = F/S_0$ (here S_0 is the cross-sectional area of the sample at t = 0) are given in Fig. 4.

In the region of t under investigation these dependencies are strictly increasing and practically independent of σ at low values of t. The higher σ_0, the faster they grow with time.

2.2.4 Dependencies of Stress and Strain Velocities upon Elastic Strain

Let us consider also the regularities common for different types of extension. Dependencies of extension σ upon elastic strain α are given in Fig. 5. Continuous lines in Fig. 5 indicate dependencies σ(α) in extension at different constant strain velocities ϰ. The higher ϰ, the higher passes the dependency σ(α). The points with maximum α

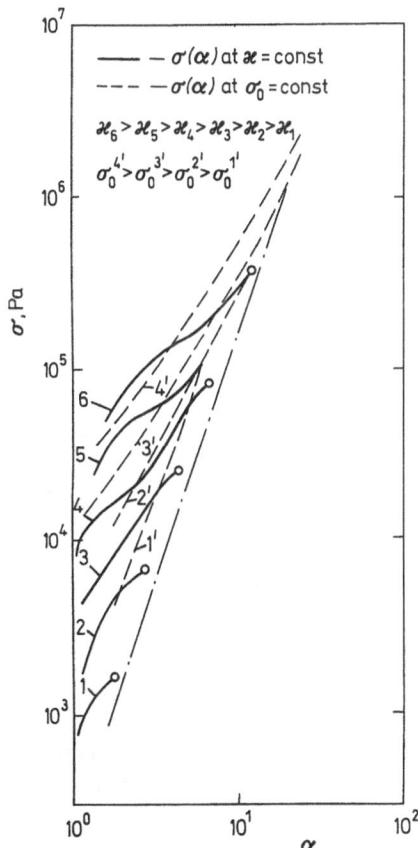

Fig. 5. Stress σ versus elastic strain α under different conditions of extension [23]

correspond to the stationary flow in cases when it has not been attained. It is clear from Fig. 5 that stress σ depends not only upon α (as it is often assumed) but also upon strain velocity $\varkappa$. The dashed lines in Fig. 5 indicate dependencies $\sigma(\alpha)$ obtained in extension at different constat forces (initial stresses σ_0). The higher σ_0, the higher passes the corresponding curve.

The comparison of dependencies $\sigma(\alpha)$ obtained under conditions of constant force and constant strain velocity indicates that the strain velocities are approximately similar at the points of their crossing.

Dependencies $e_p(\alpha)$ [23, 26, 35] under conditions of $\varkappa$ and $F = $ const had a form similar to $\sigma(\alpha)$.

The dotted line in the right part of Fig. 5 indicates the envelope curve of dependencies $\sigma(\alpha)$ which are located leftwards therefrom. There is also an envelope for curves $e_p(\alpha)$. Envelopes of dependencies $\sigma(\alpha)$ and $e_p(\alpha)$ were obtained in Refs. [23, 26] from experiments in stress relaxation. Stationary values $\sigma(\alpha)$ and $e_p(\alpha)$, in tests with $\varkappa = $ const, are located practically on these envelopes.

The existence of an envelope in dependencies $\sigma(\alpha)$ provides a possibility to impose a limitation on the unknown elastic strain α proceeding from the known tensile stress. This is important in manufacturing of fibers and films since, as it is demonstrated in Ref. [19], the strength of a polymer vitrified after extension grows with increase in strain.

Comparison of dependencies $\sigma(\alpha)$ and $e_p(\alpha)$ under different extension conditions and at stress relaxation [23, 39, 40] indicates that if in two different processes both σ and α coincide, the value e_p also coincides. Thus there is a functional dependency:

$$\sigma = \sigma(\alpha, e_p) \tag{8}$$

while, as indicated in Refs. [13, 14] at $\alpha \rightarrow 1$:

$$\sigma = 3\eta e_p \tag{9}$$

2.2.5 Stress and Strain Relaxation Processes

The existence of the functional dependency of Eq. (8) provides a possibility [41] to assume that relaxation processes develop in time similarly after different homogeneous extensions (and even after a preliminary partial relaxation of stresses), if the accumulated elastic strain α and irreversible strain velocity e_p in the medium are similar at the moment of start of relaxation. According to the above, we can take, for example, instead of parameters (α, e_p) parameters (σ, α). If the processes of stress relaxation coincide in time, the processes of "retardation" occur "automatically" similarly under the same initial conditions of (α, e_p).

2.2.6 Some Differences in the Behavior of Effective Viscosity under Extension

In Refs. [17, 18, 21] the uniaxial extension of polyethylene at a constant strain velocity was considered. Figure 6 gives the dependencies of effective viscosity $\sigma/\varkappa$ (σ is tensile stress, $\varkappa$ is strain velocity) upon time t obtained in final form in Ref. [21]. The stationary flow (i.e., a constant value of $\sigma/\varkappa$) was attained practically for all values of $\varkappa$. The higher $\varkappa$, the higher passes the respective curve. The lower curve $3\sigma_{12}/\dot{\gamma}$ (here σ_{12} is

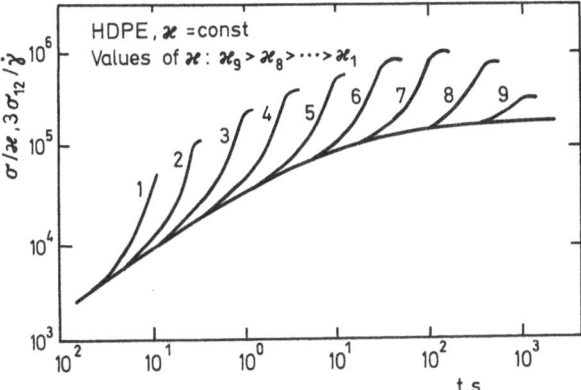

Fig. 6. Effective viscosity versus time in extension of polyethylene under conditions of constant strain velocities [21]

tangential stress, $\dot{\gamma}$ is shear velocity) represents effective viscosity determined at shear in the region of the medium's linear behavior. It should coincide at any moment t with effective viscosity $\sigma(t)\varkappa$ under extension in the linear strain region and should not depend upon the strain velocity. The relationship $\sigma(t)\varkappa$ under extension with further increase in $\varkappa$ also did not depend upon $\varkappa$ (see Fig. 6) at short flow times and coincided with $3\sigma_{12}(t)/\dot{\gamma}$. But further, with increase in t, the dependencies $\sigma(t)/\varkappa$ did not correspond to linear strain and deviated from $3\sigma_{12}(t)/\dot{\gamma}$.

The major difference between dependencies of $\sigma/\varkappa$ upon t, in case of polyethylene, and the similar curves in case of polyisobutylene (see Fig. 4) was that even at very low values of tensile strain velocities $\varkappa$ in case of polyethylene, the flow failed to remain linear during the entire time interval of the experiment. Dependencies of effective viscosity $\sigma(t)/\varkappa$ at low values of $\varkappa$ reached a constant value (stationary flow) exceeding 3η during times $\tilde{t}$, significantly exceeding the similar time under shear. With further increase in strain velocity $\varkappa$, the value t started to decrease in the same way as it was in experiments with polyisobutylene (PIB).

The effective viscosity under stationary flow, with increasing $\varkappa$, was growing significantly at first and then reduced. Reduction in the viscosity of PIB [23, 26] also was not observed, apparently, because of limitations of instrumentation.

Some regularities, similar to Refs. [17, 18, 21], of viscosity variation in time under conditions of $\varkappa$ = const were observed by the authors of Refs. [23, 24, 32–34] in extension of polyethylene and butadiene rubber (BR). Note that in Ref. [34] the linear region of strain reaching the stationary flow was attained in extension of butadiene rubber. With further step-wise increase of $\varkappa$ the effective viscosity grew and the time during which the stationary flow was attained increased significantly. References [23, 24] will be discussed in more detail below.

Thus, on the basis of facts discussed in this section, we may conclude that effective viscosity $\sigma/\varkappa$, in extension of some polymers, passes the maximum with increasing strain velocity $\varkappa$. The first time t measured reliably in the nonlinear strain region exceeds significantly the respective time θ in the linear area. Dependency $\varkappa$ decreases with further increase in t($\varkappa$).

2.2.7 Retardation of Polymer Fluid Flow under Great Elastic Strains

This section will deal with suppression of flow in extension of molten polymers in the region of significant elastic strains [21, 24]. The study of polyisobutylene II-20 [23, 35] failed to reveal such phenomena (the velocity of irreversible strain $e_p = d(\ln \beta)/dt$ increased strictly with time). Retardation of polymer fluid flow is considered on the example of homogeneous extension at constant strain velocity and force. Most experiments were carried out with commercial low-density polyethylene (LDPE) with molecular weight MW $\sim 10^5$. Figure 7 gives experimental dependencies of tensile force F/S_0 and irreversible strain $\ln \beta$ ($\beta = \varepsilon/\alpha$) upon time t at different

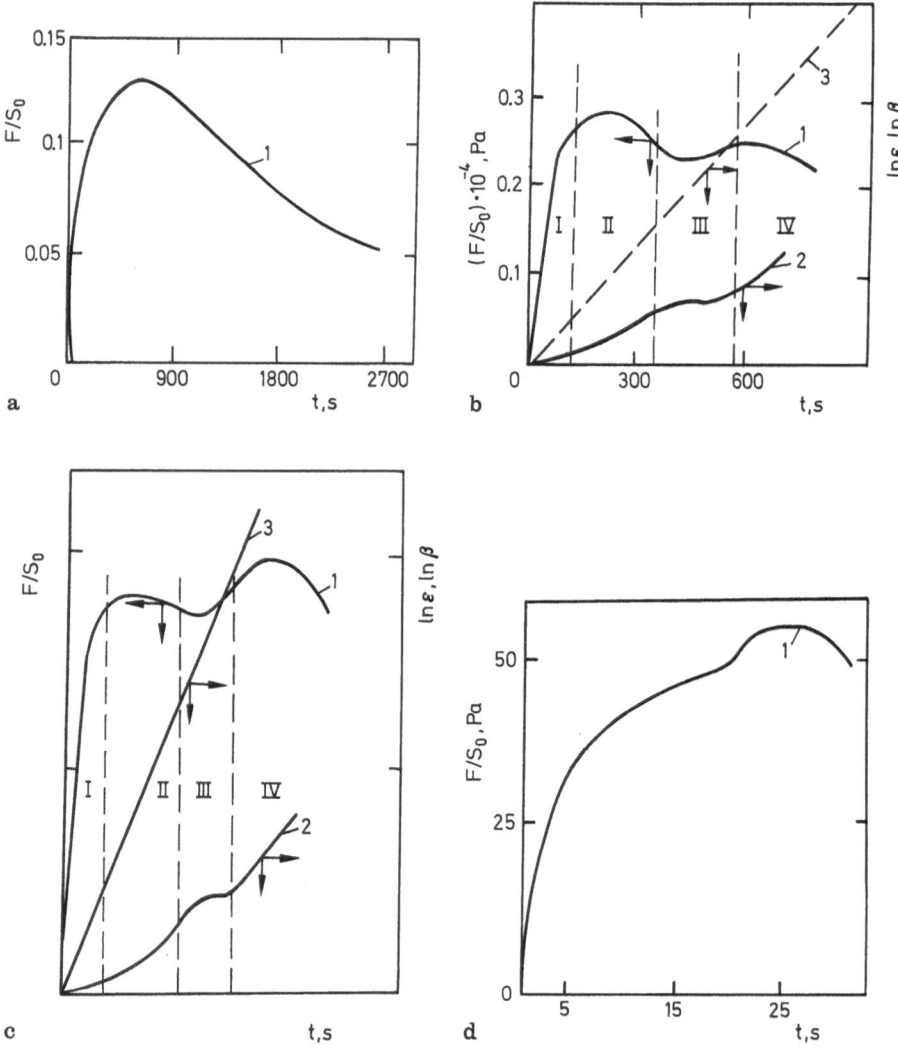

Fig 7 a–d. Tensile force F, total ($\ln \varepsilon$) and irreversible ($\ln \beta$) strains versus time t at different strain velocities [23, 24] (see text for explanation)

strain velocities $\varkappa$. The straight line indicates also dependencies of total strain $\ln \varepsilon = \varkappa t$. Illustrations 1, 2, and 3 in Fig. 7 correspond to dependency of F, $\ln \beta$ and $\ln \varepsilon$ upon time t.

At low values of $\varkappa$, normally (see Fig. 7), one maximum in dependency F(t) was observed, as it was in case of polyisobutylene; at high values of $\varkappa$ there were already two maxima. In this case (see Fig. 7c, d) four regions of strain were observed (see also Sect. 1 of this review).

Region I starting from the moment of time t = 0 is characterized by small difference between elastic strain $\ln \alpha$ and total strain $\ln \varepsilon$. The time of straining in Region 1 is significantly shorter than the relaxation time θ (which was determined by the time during which stationary flow was attained in the linear strain region) and the polymer is strained similar to cross-linked rubber. The force grows in this region.

In Region II the flow of polymer medium develops (the velocity of irreversible strain e_p increases). In this region the force F(t) passes the maximum and starts to decrease.

Region III begins at the point of inflection of dependency F(t) in its decreasing section. This point corresponds to the point of inflection of the dependency $\ln\beta(t)$ which increases. In this region the velocity of irreversible strain e_p starts to decrease with increase in t and approaches zero ($\ln \beta$ = const), i.e., the irreversible flow in this section is suppressed and the medium starts to strain similarly to the cross-linked rubber. In this case the force in section $e_p \approx 0$ grows again which is also characteristic of the deformation of solid bodes under conditions of $\varkappa$ = const.

In Region IV the repeat flow develops again ($e_p > 0$).

Note that in all four regions the polymer remains similarly transparent and is strained uniformly.

At even higher values of $\varkappa$ (see Fig. 7d) the minimum in the dependency F(t) disappears and the second growth begins immediately after the first growth of the force (Region II is practically absent). Thus the region of elastic strain, at times shorter than relaxation time θ, coalesces with the area of repeat "hardening".

In case of extension of polyisobutylene (see Fig. 1) the "hardening" effects were not observed in the investigated region of strain velocities at room temperature. Dependencies of F(t) had one maximum (as in case of polyethylene at low values of $\varkappa$) and then decreased strictly while the velocity of irreversible strain was strictly increasing ($e_p = d(\ln \beta)/dt$).

Note that despite a certain similarity between dependencies of $\sigma/\varkappa$ upon t obtained in extension in Refs. [21] and [23, 24] but, in contrast to [23, 24], in Refs. [17, 18, 21] the effects on the prestationary region of extension discussed in this section were expressed apparently weaker.

Two maxima in the dependency F(t) were observed also in extension of molten high-density polyethylene [24]. In case of the high-density polyethylene a break of the sample preceded by formation of a neck was observed after the secondary growth of F(t). The repeat growth of force following directly the first one (similarly to Fig. 7) was observed also in noncrystallizing polymer, molten polystyrene (PS). Elastic strain was not measured in extension of the two above mentioned polymers. If "hardening" is considered from the point of view of the repeat growth of force, we may state that this property, in principle, is characteristic of polymer fluids of different nature.

Thus, after the secondary hardening, either the secondary flow develops or the materials break. Note that breaks, in principle, can also take place in the region of strain times shorter than the relaxation time θ, and may not characterize the region of the secondary "hardening".

Let us consider now the effect of flow retardation in extension under constant force [25]. The above-mentioned experiments were carried out with low-density polyethylene at 125 °C and with polyisobutylene II-20 at 44 °C. It should be noted right away that at the above-specified temperatures these melts have approximately similar characteristics under shear strain: maximum viscosity is $\eta \approx 3 \times 10^5$ Pa s and relaxation time is $\theta \sim 10^2$ s. Flow curves in the investigated region of shear strain velocities also did not differ significantly from one another.

The results of experiments are given in Fig. 8. These are dependencies $\ln \varepsilon(1)$,

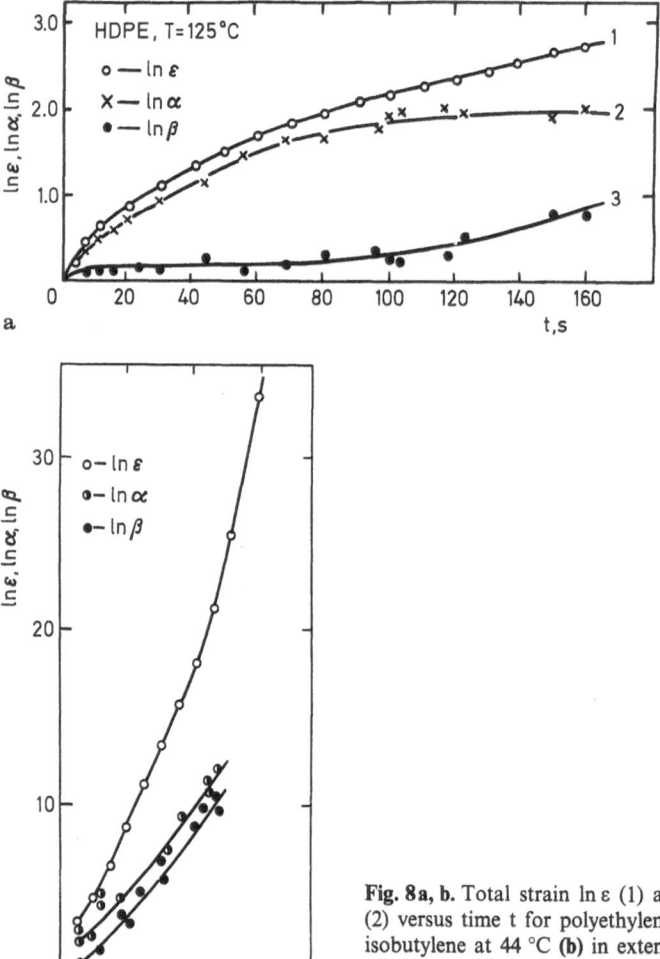

Fig. 8a, b. Total strain $\ln \varepsilon$ (1) and irreversible strain $\ln \beta$ (2) versus time t for polyethylene at 125 °C (**a**) and polyisobutylene at 44 °C (**b**) in extension under constant-force conditions ($\sigma = 1 \times 10^4$ Pa) [25]

$\ln \alpha(2)$, $\ln \beta(3)$ upon time of extension for low-density polyethylene (Fig. 8a) and lyisobutylene (Fig. 8b) at the same σ_0, where $\sigma_0 = F/S_0$ (F is the tensile force, S_0 is the area of initial cross section of the sample).

Despite the proximity of flow curves, as well as values η and θ for these polymers (values η_0 and θ are usually subject to dimensionless representation, normalization of σ_0 and t), the time during which the maximum possible strain is attained $\ln \varepsilon_{max} = 2.8$ for polyethylene and exceeds that of polyisobutylene at the same σ_0 by factor of 6. In this case the dependency $\ln \varepsilon(t)$ in the region of measurement for polyisobutylene is characterized by increasing strain velocity $\varkappa = d(\ln \varepsilon)/dt$ in contrast to which $\varkappa(t)$ decreases strictly in low-density polyethylene within a significant section of ε and becomes approximately constant at high values of t.

2.3 Theoretical Description of Extension in Molten Polydisperse Polymers

Numerous works aimed at theoretical description of the extension experiments (see, e.g. [42]) failed to produce satisfactory results. It should be noted that description of shear experiments was adequate in many works (see, e.g. [43–45]).

Finally, in Ref. [46] a satisfactory description of extension experiments and model calculations [47] was provided.

The basis for the above-mentioned model [47] was provided by Maxwell's nonlinear model obtained in general form in Ref. [48]. Here the total strain was divided into irreversible strain and elastic strain λ. Stress σ and velocity of irreversible strain e_p were determined from the elastic strain. In Ref. [48] a number of functions $\sigma(\lambda)$ and $e_p(\lambda)$ were defined more specifically. Beside that, Maxwell's nonlinear models were connected in parallel. Note that in case of one Maxwell's element $\lambda = \alpha$ [23], but in case of several elements connected in parallel this is not true and α is determined from the solution of the respective problem. In case of the uniaxial extension the model of Ref. [47] takes the following form:

$$\frac{\theta_k}{\lambda_k} \frac{d\lambda_k}{dt} + \exp(m_k W_k) \frac{(\lambda_k + 1)(\lambda_k^3 - 1)}{6\lambda_k^2} = \varkappa \theta_k$$

$$\sigma = 4 \sum_{k=1}^{N} \frac{\mu_k}{n_k} (\lambda_k^{n_k} - \lambda_k^{-n_k/2}) + 3S\eta\varkappa$$

$$W_k = \frac{4M_k}{n_k^2} (\lambda_k^{n_k} + 2\lambda_k^{-n_k/2} - 3) \tag{10}$$

where K is the number of Maxwell's elements, N is their quantity, θ_k, and μ_k are, respectively, relaxation time and modulus (constants of the medium); constant n_k is the power of elastic potential W_k; λ_k is elastic strain; σ is tensile stress; t is time; $\varkappa$ is strain velocity of uniaxial extension [see Eq. (4)].

Note that the model given in Ref. [47] takes into account the distortion of potential barriers in activation flow under the action of stress. Relaxation time is $\theta_k^{\varkappa} = \theta_k \exp$ · $(-mW_k)$ (here $m_k = \dot{\gamma}_k/RT$ where $\dot{\gamma}_k$ is constant, R is gas constant, T is temperature). Normally, two Maxwell's elements (N = 2) and one viscous element with viscosity $S\eta_0$ [here S is constant $(0 < S < 1)$, and η_0 is the maximum Newtonian viscosity $\eta_0(1 - S) = \sum_{k=1}^{N} \eta_k$, $\eta_k = 2\mu_k\theta_k$] are connected in parallel.

The following limitations are imposed on the constants of the medium [47]:

$$\theta_k \gg \theta_{k+1}, \mu_k \ll \mu_{k+1}, \eta_k\theta_k \gg \eta_{k+1}\theta_{k+1}$$
$$4 \gg n_k \gg 2$$

In case the distortion produced by the stress of potential barriers is insignificant $(n_k W_k \ll 1)$ the following asymptotic dependencies $(\Gamma_k = \theta_k\varkappa)$ were obtained under stationary flow:

$$\max \Gamma_k \ll 1G \approx 3\eta\varkappa$$

$$\min \Gamma_k \gg 1 \sigma \approx 4 \sum_{k=1}^{N} \left(\frac{\mu_k}{n_k}\right)(6\Gamma_k)^{n_k/2} + 3S\eta\varkappa \tag{11}$$

At $n_k > 2$ the viscosity $\sigma/\varkappa$ in Maxwell's element grows unlimited with increase of $\varkappa$. When value $m_k W_k$ is not small, the activation mechanism makes the viscosity function pass under stationary flow from $\varkappa$ via a maximum, which was observed in experiments described in Ref. [21].

The model under consideration describes adequately [46], from the quantitative point of view, the experiments in extension, the results of which are given in Fig. 3–8. In this case, if the following inequality holds true (at all k) for viscosities $\eta_k = 2\mu_k\theta_k$: $\eta_k \gtrsim \eta_{k+1}$ there is only one maximum on the dependency of force F upon time t in extension under conditions $\varkappa = $ const, and time during which the stationary flow is attained strictly decreases with increase of $\varkappa$ (see Fig. 3).

Should we assume that there is a significant relaxation time θ_1 with small weight [46, 47] in the system (see below for the physical mechanism of existence of such θ_1 in monodisperse polymers):

$$\eta_1 \ll \sum_{k=2}^{N} \eta_k, \qquad \eta_2 \gtrsim \eta_3 \gtrsim ... \gtrsim \eta_N,$$

this relaxation mechanism in the linear strain region is insignificant. It shall not affect the flow curve since effective viscosity decreases in shear. However, this mechanism may become significant and even determinant in extension, when effective viscosity increases. Its existence [46, 47] results in the 'elongation' of time during which the stationary flow is attained (Fig. 6, 7) since $\theta_1 \gg \theta_2$. Time θ_2 is approximately equal to the stationary-flow time in the linear flow region. The existence of a long relaxation time with small weight results in appearance of the second maximum in dependencies F(t) (Fig. 7) since, along with the Deborah's number $D_2 = t/\theta_2$, there is a new criterion $D_1 = t/\theta_1$ (t is time).

In Refs. [46, 47] a quantitative correspondence between the results of analysis and experiments carried out with molten polyethylene in Ref. [24] was obtained.

In extension under conditions of $F = $ const a large-scale relaxation mechanism with small weight may result in retardation of the flow compared to a system where there is no such effect. In this case flow curves may, practically, not differ from one another.

3 Extension of Molten Polymers and Molecular-Kinetic Theories

Numerous phenomenological models suggested for description of viscoelastic properties of molten polymers illustrate the complexity of an adequate description of
rheological behavior of these media under various external effects, especially under
extension. Therefore, the development of physical knowledge about relaxation mechanisms conditioning nonlinear effects observed in extension of molten polymers is
extremely important. Development of molecular-kinetic theories allows to improve
phenomenological models of such polymer media. In recent years the theory of viscoelasticity of molten polymers was progressing exactly on the basis of molecular-kinetic
knowledge.

Viscoelastic properties of molten polymers conditioning the major regularities of
polymer extension are usually explained within the framework of the network concept
according to which the interaction of polymer molecules is localized in individual,
spaced rather far apart, engagement nodes. The early network theories were developed
by Green and Tobolsky [49] and stemmed from successfull network theories of rubber
elasticity. These theories were elaborated more fully in works by Lodge [50] and Yamamoto [51]. The major elasticity. These theories is their simplicity. However, they
have a serious drawback: the absence of molecular weight in the theory.

From this point of view the Doi-Edvards' reptation theory can be regarded as the
most perfect network theory [52]. In a molten polymer, macromolecules can not move
notable in lateral direction since that is impeded by other polymer chains. This circum-
stance in the Doi-Edvards theory is taken into account by means of introduction of a

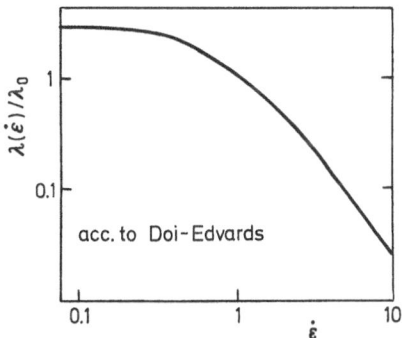

Fig. 9. Viscosity in extension according to Doi-
Edwards theory

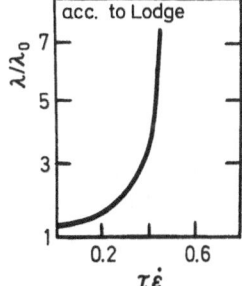

Fig. 10. Longitudinal viscosity in molten polymers according to
Lodge's theory

tubular element limiting the lateral fluidity of the polymer chain. Movements of a polymer chain resembling "creeping" along such a tube have been called reptations. According to the Doi-Edvards theory, the longitudinal viscosity strictly decreases with increase in elongation velocity (Fig. 9). Note that Lodge's network theory of molten polymers [53] leads to a conclusion about strictly increasing longitudinal viscosity with increase in elongation velocity (Fig. 10).

In description of effects observed in extension of molten polymers, the determinant is the phenomenon of anisotropy of the mobility of macromolecules. In the Doi-Edvards reptation theory the anisotropy of the mobility of macromolecules is taken into account topologically by means of placing a macromolecule into a certain hypothetical tube. In this case large-scale movements are allowed only along the macromolecule and are totally inhibited in the lateral direction. This, indeed, is a limiting case of mobility anisotropy.

It has been demonstrated [54] that the dependency of viscosity upon elongation velocity varies significantly in its character with alteration of the lateral (transverse) mobility of macromolecules. In case there is no lateral mobility of macromolecules, the longitudinal viscosity strictly decreases under tension (see Fig. 11) which corresponds to prognostications proceeding from the Doi-Edvards theory. In case there is a lateral mobility of macromolecules, an insignificant maximum or monotonous increase in longitudinal viscosity of the medium with increase in extension velocity is observed.

The reptation theory of viscoelasticity developed by Doi & Edvards has paractically predetermined the appearance of Curtiss-Bird's reptation theory [55]. The latter is constructed on the basis of general kinetic theory of polymer fluids in phase space. According to the Curtiss-Bird theory, the longitudinal viscosity may increase only by a factor of two compared to the initial value. Note that a more significant increase in longitudinal viscosity was observed in experiments.

In the molecular-kinetic theory of molten polymers there is another approach to the problem which does not employ the notion of engagements. It is based on the analysis of dynamics of individual macromolecules in the medium of their like. In this case the real environment of macromolecules is substituted by a certain averaged continuous medium. The major problem of this approach is to determine the character of this me-

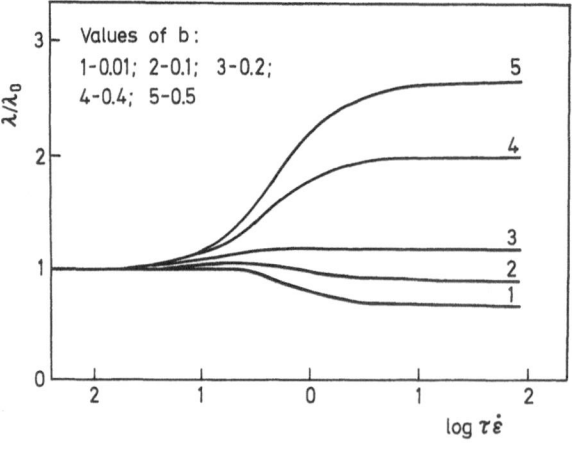

Fig. 11. Longitudinal viscosity at different values of the parameter of lateral mobility of macromolecules b

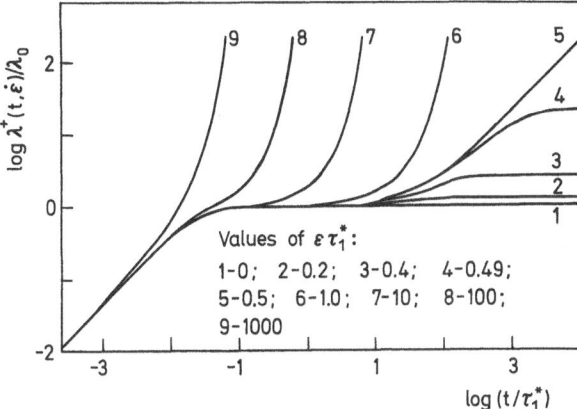

Values of $\varepsilon\tau_1^*$:
1-0; 2-0.2; 3-0.4; 4-0.49;
5-0.5; 6-1.0; 7-10; 8-100;
9-1000

Fig. 12. Time function of viscosity in uniaxial extension for different dimensionless elongation velocities $\dot{\varepsilon}\tau_1^*$

dium. Modelling of macromolecules of the surrounded relaxing (viscoelastic) continuous medium is of special interest [56, 57]. According to this point of view, polymer macromolecules move in a medium which relaxes with a certain characteristic relaxation time determined from the condition of self-consistence. Figure 12 gives viscosity n extension at constant elongation velocity $\dot{\varepsilon}$ of a molten polymer [$\alpha(t, \dot{\varepsilon})$] as a function of extension time t according to the above-mentioned theory of microviscoelasticity. The transient longitudinal viscosity $\lambda(t, \dot{\varepsilon})$ gives a very clear idea of relaxation processes which take place in a molten polymer under extension.

In the region of linear behaviour of a melt the viscosity strictly increases up to the stationary value $\lambda_0 = 3\eta$ of the linear region; the transfer to a stationary flow under extension is determined by relaxation processes with characteristic relaxation time $\theta \sim (MM)^{3,4}$. A clear manifestation of that is a very drastic transfer to the stationary value of λ_0, which should be expected in case of ideally monodisperse polymers. Note that the duration of the transfer of longitudinal viscosity in the respective region of linear viscoelasticity to the stationary value allows to judge about the polydispersity of a polymer. Polymers with a wide molecular-weight distribution (MWD) are characterized by a slower transfer to stationary flow. Therefore, stresses in a polydisperse polymer in the beginning of extension are lower than in the case of a polymer with relatively narrow MWD (quasimonodisperse polymer).

In the area of nonlinear behavior at relatively low extension velocities, the time of transfer to the stationary flow by far exceeds the relaxation time $\theta \sim (MM)^{3,4}$. These phenomena have a characteristic relaxation time $\theta \sim (MM)^{4,4}$, and are connected with superslow relaxation processes. The above phenomena are observed in polymers with a high molecular weight [60].

At high extension velocities the viscosity does not reach the stationary value but grows unrestrictedly with time. In this case there is no stationary process of uniform extension. This, however, does not mean at all that a molten polymer can not be subjected to extension at high elongation velocities. In this case the longitudinal viscosity never reaches the ultimate value. It grows unrestrictedly with time. Nevertheless, in any real experiment in extension one should not expect infinite values of viscosity. The duration of any flow under extension is restricted practically always since great elongation is attained quickly, and the sample is destroyed. The higher the elongation velocity, the shorter the permissible duration of the experiment. In the region of high elongation velocity the only matter of practical interest is the character of increase in the longitudinal viscosity at the stage of the transient process. Increase in viscosity at the initial stage of extension does not depend upon the molecular weight of a polymer and upon the elongation velocity. It is determined by the value of elasticity modulus on the plateau G_p:

$$\lambda^*(t, \dot{\varepsilon}) = 3\,G_p t \tag{12}$$

Dependencies of longitudinal viscosity upon time at different extension velocities coincide up to the start of deviation from linear behavior. The higher the extension velocity, the earlier deviation is observed. An interesting fact is that in all cases the deviation from linearity is observed at one and the same critical strain independent of the molecular weight of the polymer and temperature. The value of critical strain depends only upon the nature of the polymer and lies within the limits of $0.4 - 1$.

According to the relaxing interaction theory [57], deviation from linear behavior is determined by superslow relaxation processes with the characteristic relaxation time $\theta \sim (MM)^{4,4}$. This deviation is insignificant in polymers with a narrow molecular-weight distribution [59] and can be neglected in a first approximation [3]. In this case the longitudinal viscosity remains practically constant and equal to $\lambda_0 = 3\eta_0$ up to the destruction of the sample. Note that in the region of extension velocities corresponding to nonlinear conditions, the constant values of longitudinal viscosity of polymers with a narrow molecular-weight distribution should be considered as quasistationary values. In this region the true stationary flow conditions are not attained and extension is terminated by a break of the polymer practically immediately after deviation from the linear dependency.

The superslow relaxation processes in polymers with a high molecular weight have been revealed for the first time in Ref. [60] which studied specific features of rheological behavior of monodisperse polymers with a high molecular weight under harmonic effects. They are an extremely important characteristic for technical application of high polymers since they determine the character of polymer behavior under effects continuing for a very long time or at very low frequencies. They also characterize the specific behavior of molten polymers at high extension velocities at the final stage of drawing. It is of major importance for polymer processing since it sets a limit for possibilities of intensification of molding by means of increase in the strain velocity during extension.

4 Some Technological Applications in Thermoplastic Processing

4.1 General Aspects

As mentioned in the beginning of this review (see Sect. 1), besides the theoretical importance of modelling and experiments in extension of molten polymers, there is an increasing interest in this field of rheology and mechanics of viscoelastic fluids from the technological point of view. This is connected with a wide spectrum of applied problems, the solution of which is based on data on melt extension. Below we shall discuss some important aspects of this problem and recent achievements in application of melt extension as an efficient technological procedure.

Processing of materials — a set of technologies including preparation of materials and their transformation into samples with preset configuration and dimensions or semifinished products and products with required operating characteristics — depends primarily on the so-called technological properties of polymers and compounds based thereon. The term "technological properties" is very complicated and versatile despite its apparent simplicity. It covers a set of numerous characteristics of the properties of materials the list of which depends upon specific objective of research, design and development, and technological problems. Classification of the basic technological properties, their rather adequate description and interconnection with processing parameters can be found, for example, in a recent review [2] and many articles, e.g. [61]. All technological properties of materials (thermophysical and structural characteristics, looseness, content of volatile materials, heat stability, etc.) can not be discussed within the scope of this publication so we limit ourselves to technological estimations of the rheological properties of polymers in processing of these materials.

4.2 Study and Estimation of Technological Properties of Polymers

Leading manufacturers in the USSR and Western companies classify each brand of polymer primarily according to the type of manufactured products (for example, extrusion brands are classified into film brands, tube brands, sheet brands, etc.). Normally, the melt flow index (MFI) determined according to GOST 11 645-73 (USSR) and CMEA Standard 896-78 or the melt index, equivalent thereto, according to US standards (ASTM 1238-73), ISO (1133-75), FRG (DIN 53735), or Great Britain (BS) is taken as the main technological characteristic of rheological (or viscous, to be more precise) properties. Conditions of experiments carried out to estimate MFI with the help of extrusion plastomers made by different companies; such as ИИРТ-М and ИИРТ-А made by NPO Himavtomatik (USSR) [62], SCHMELZINDEXER made by Gottfert Mestechnik (FRG), MELT INDEX made by Chori (Japan) and by Ceast (Italy) are described in respective standards, manufacturer's booklets, and in reference literature. Estimation of materials in terms of MFI provides usually a basis for classification of thermoplastics (low-density polyethylene, high-density polyethylene, polypropylene, polystyrene, polyethylene terephthalate, etc.) in terms of the major recommended type of processing. The arbitrary nature of such assessment is clear since, by changing temperature and shear stress (load) one can vary within a significant range the viscosity of the melt and, consequently, the effective value of MFI if that is not impeded by thermal destruction of the polymer. For a number of comparatively new processes, such as manufacturing of thin and ultrathin oriented films, film threads, fibrillated fibers, etc., polymers can be used which are usually classified, in terms of MFI, not as extrusion but as casting brands; in this case the high-elastic properties of melts are also very important. Finally, the materials characterized by similar values of MFI can posses different rheological and, consequently, technological properties, especially in case of polymer batches produced by different manufacturers. This is accounted for by the fact that MFI reflects a certain mean value of the molecular weight (MW) which may correspond to materials different in terms of other characteristics, such as molecular-weight distribution and ramification of macromolecules in samples (batches) of polymers. Indeed, variation of molecular-weight distribution and polymer structure allows to alter significantly rheological (viscous and high-elastic) properties of the material and these changes can not be reflected by one standardized MFI characteristic under shear strain. The above discussion will be exemplified below. The most radical changes in rheological properties take place in transition from polymers with a narrow molecular-weight distribution (i.e., from practically monodisperse polymers) to sample with broad molecular-weight distribution (to which belongs the overwhelming majority of commercial materials). Since the value of MFI corresponds to only one point on the flow curve, on the dependency $\tau(\dot{\gamma})$; (here τ and $\dot{\gamma}$ are, respectively, stress and velocity of shear), it is clear that estimation of technological properties by MFI does not allow one to prognosticate the so drastic difference in the behavior of mono- and polydisperse polymers in processing, and does not reflect the form of the flow curve. Apparently, new standard procedures should be developed to provide a background for analysis of the character of molecular-weight distribution. The simplest method of such estimation is the introduction of a quantitative measure of description of the flow curve form reflecting the character of molecular-weight distribution. Although the connection between the flow curve form and molecular-weight distribution has not been established unambiguously so far, there is no doubt that it does exist. Numerous attempts were made to characterize the polymer raw materials not by one point on the flow curve but by at least two points (see, e.g. [2, 63]). The easiest way to do that is to use the relationship of two MFI values measured under different loads $(J = MFI_1/MFI_2)$. Value J depends strongly on the polydispersity of the material and it is a useful, to some extent, characteristic of technological properties, at least in polymers with a comparatively narrow molecular-weight distribution. However, even a coincidence of materials by two values MFI_1 and MFI_2 determined under significantly different shear stresses and by full flow curves (determination of which is rather 'labor consuming' and 'unprofitable' for conditions of commercial production) does not guarantee the absence of technological 'troubles'. In real experience it is often necessary to change process conditions in transit from one material to another in order to avoid different thickness of products, the spread of physical and mechanical characteristics, etc. (the causes of that will be explained below).

A process engineer can obtain useful information from the data on viscous flow activation energy (E) [2, 3, 7] but it is more important not in certification of different batches of one and the same material but in comparison of polymers of different nature, since value E, as well as the above-mentioned characteristics, is determined under shear strain and is relatively in sensitive to insignificant changes in polymer structure. The use of other rheological characteristics measured under shear strain (e.g. normal stress difference) and under vibration strain conditions (elasticity modulus, loss modulus, etc.) to estimate the raw materials requires unique equipment which is hardly acceptable for certification of the polymer raw materials. This, apparently, accounts for the fact that publications dealing with this matter are so scarce. We may also mention that certification of rubbers was performed with the help of a vibrorheometer [64], quality control of high-molecular polymers was performed with the help of a rotary viscosimeter [65], and dynamic mechanical characteristics of thermoplastics were measured with the help of original mechanical spectrometers RHEOMETRICS (USA) and DXΠ (USSR) [66].

Much work has been done within the framework of the international programme JUPAC by 14 laboratories of different companies to reveal a characteristic of polymer material correlating with specific features of its molding into sleeve film[67]. The studies were conducted with two samples of low-density polyethylene and two samples of high-density polyethylene. The samples had minor differences in molecular-weight distribution and ramification characterized by the content of different side branches.

Experimentators managed to produce thin films from all polymers under investigation but their thickness differed with certain regularities despite the fact that the films were produced at different laboratories and on different equipment. The shape of sleeve and temperature profile over sleeve length also differed significantly, as well as the specific impact elasticity of films both along and transverse to the direction of extrusion.

To identify characteristics which determine the behavior of materials in processing, a large number of characteristics of the raw material was estimated: 1) thermal stability on the basis of MFI variation and flow through capillary viscosimeter depending on the time of maturing; 2) behavior under crystallization; 3) viscosity under stationary shear flow, flow curves; 4) first difference of normal stresses; 5) modulus of elasticity and modulus of losses; 6) relaxation of stresses after shear strain; 7) relaxation modulus and reversible strain after shear flow; 8) pressure drop under convergent flow; 9) double refraction with the help of a slot capillary; 10) swelling of extrudate; 11) total, reversible, and irreversible strains under longitudinal strain under constant stress drawing conditions ($\sigma = $ const); 12) longitudinal viscosity in stationary longitudinal flow under conditions of $\sigma = $ const; 13) critical ratio at drawing from a capillary.

Differences between materials were detected in all of the employed methods (1—13) while in the comprehensive work cited above it has been noted that elastic characteristics of materials are more sensitive to minor changes in the molecular structure than the viscous ones.

Presently, extensive theoretical and experimental data have been accumulated. These data indicate that information on technological properties of polymer materials, very important for processing, can be obtained if rheological methods based not on the shear flow but on tensile strain of the melts are employed. Therefore it is reasonable to make an emphasis on method 13, more so because differences in the molecular structure of polymers under drawing from a capillary are more distinct. Experiments, normally, include the following[67]: Rolls with a riffled surface connected with force transducer (force meter) are installed under the capillary of a viscosimeter or laboratory extruder (extrusiometer). Extrudate is passed there through, rolls are rotated with a constant acceleration which can be varied. Time alterations of tensile effort are recorded. The plant imitates a real drawing of a melt with regard to prestraining of the polymer in the molding tool (capillary) and nonisothermicity of the process and intensifies the process in order to study break characteristics of the jet. It is possible to study three simultaneous processes every one of which, in its turn, depends on a number of variables, for example, upon prestrain of the polymer in the extrusiometer or reservoir of the capillary viscosimeter, on the dimensions of the capillary, the length if the heated zone after the capillary, the elevation of rolls, their size and geometry, ambient temperature, and even on the motion of ambient air (draught).

Thus, the 13th method being most sensitive to existing differences in the structure of polymers of the same nature, it could hardly be recommended for standardization since in this case at least 7 parameteres should be specified [67]; these parameters would need to be maintained at a constant level.

Prokunin, Sevruk, and Fridman [37, 68, 69] have suggested an additional characteristic of rheological and technological properties of thermoplastics used as raw materials in such processes where an important stage is the longitudinal deformation of melts, for example, to produce films, fibers, flat threads, thermal drawing of sheet blanks, etc.

In real processes a melt is drawn from the molding tool (under laboratory conditions it is a capillary) by rolls rotating at a constant angular velocity. The result of this is inhomogeneous extrusion of the polymer. Its cross section varies over the length of the extrudate, and the form of extrudate remains constant since the material is fed from the capillary at a constant speed and it is removed also at a constant but higher speed. Thus, the extrudate is under the effect of a constant tensile force. It is difficult to study the mechanics of such a process and physical-mechanical changes which take place in the material under laboratory conditions and more so under conditions of commercial production since the flow of extrudate, as mentioned above, changes. The process develops under nonisothermic conditions which are difficult to maintain at a constant level, and dimensions of the extrudate are greatly affected by the preceding shear strain in the capillary.

To derive a directly rheological characteristic of the material under investigation from these simultaneous processes, the homogeneous extension (i.e. the cross section of samples did not vary over length) was studied under isothermic conditions [68, 69]. A relationship was found connecting the extrudate dimensions changing with time under homogeneous longitudinal strain at constant-force conditions with extrudate dimensions changing along the longitudinal coordinate in extrusion at the same conditions

$$z = v \int_0^t \varepsilon(t) \, dt \tag{13}$$

where z is the distance from the end of the capillary, v the average velocity of the extrudate at outlet from the capillary; $\varepsilon(t) = r_0^2/r^2(t) = l(t)/l_0$ is total strain; r_0, $r(t)$ is the initial and current radius of extrudate at inhomogeneous extension; l_0 and $l(t)$ are initial and current length of the sample.

This relationship allowed to model the process of inhomogeneous drawing of a melt from a capillary by a homogeneous extension [68-71]. The reliability of Eq. (13) has been verified [70]; it was demonstrated by the example of polyisobutylene that the process of swelling and inhomogeneous drawing (with regard to jet swelling) can be simulated by its homogeneous extension with different weights: one weight can be used to simulate the prestrain of the melt in the capillary, the other can simulate the subsequent drawing of extrudate.

Figure 13 gives profiles of extrudate swelling after it left the molding tool and drawn at a constant force indicated by points (obtained photographically); the lines indicate extrudate profiles obtained by calculations on the basis of Eq. (13).

Therefore it is advisable [69] to consider the method of homogeneous melt extension only at fixed constant force for future certification of raw materials.

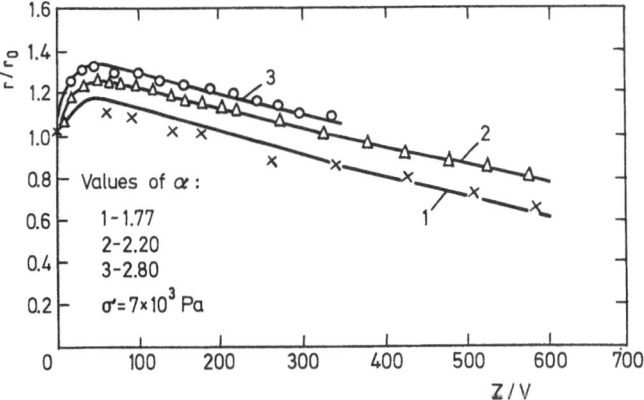

Fig. 13. Profile (r/r_e) of swelling and drawn from capillary extrudate versus ratio of the distance from the end of the capillary (z) and the average velocity of polymer leaving the capillary (V). The lines correspond to data for homogeneous extension calculated by (13) and points correspond to experimental data obtained photographically

It should be noted that the suggested method is highly sensitive to minor changes in the polymer structure of polystyrene PS) [72]. Figure 14 gives alteration of the total strain $\ln \varepsilon$ with time at extension under constant-force conditions at 140 °C. Polystyrene of two brands was used: PS-168 N, BASF-product (FRG) and ПC-151 (USSR) used for manufacturing of flat films and threads. A significant difference in the behavior of these brands under longitudinal strain is visible, although it is not detected at shear (see Fig. 15). Investigations have revealed [72] that the observed divergence of the properties of samples at extension is caused by differences in molecular-weight distribution of the investigated brands of polystyrene (see Fig. 16).

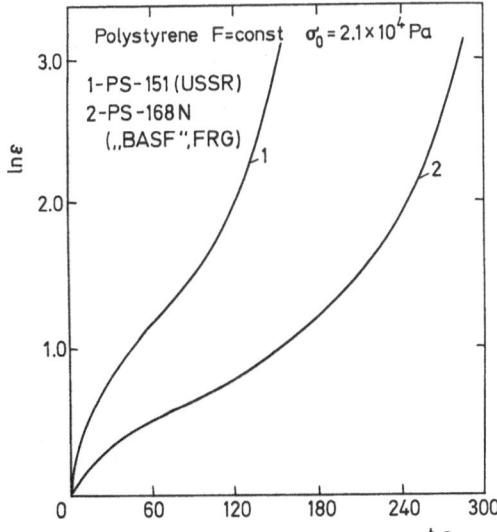

Fig. 14. Alteration of total strain ($\ln \varepsilon$) with time (t) in extension at constant force (F = const, $\sigma_0 = 2.1 \times 10^4$ Pa) in case of polystyrene of two different brands manufactured in the USSR and in the FRG

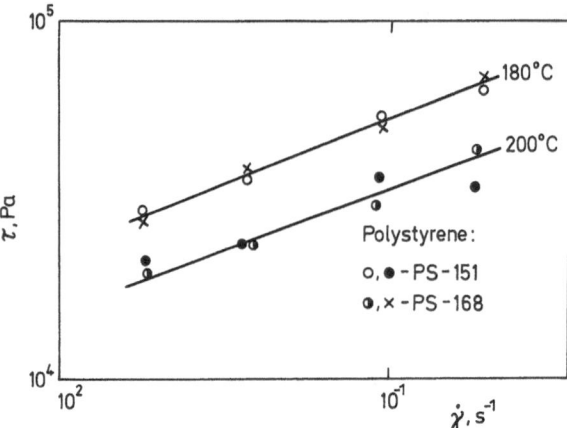

Fig. 15. Stress τ versus shear velocity ($\dot{\gamma}$) in polystyrene of two different brands (same as in Fig. 14)

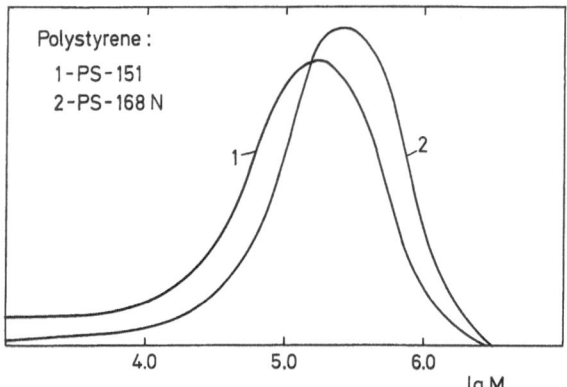

Fig. 16. Characteristics of molecular-weight distribution in polystyrene of two different brands (same as in Fig. 14 and 15)

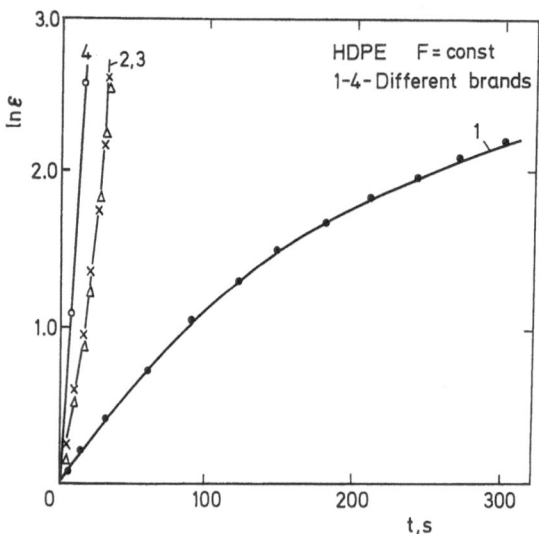

Fig. 17. Variation of total strain (ln ε) with time (t) in extension of high-density polyethylene of four different brands under conditions of constant force (F = const)

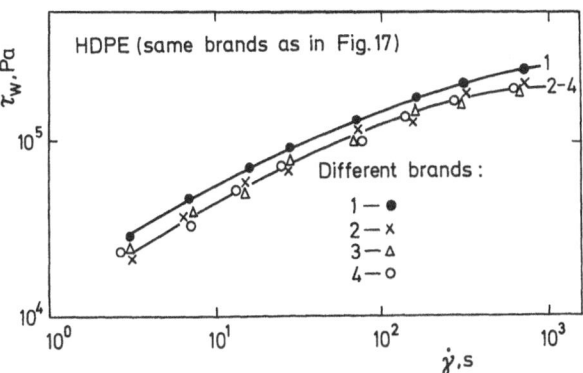

Fig. 18. Stress τ versus shear velocity $\dot{\gamma}$ (flow curves) in different brands of high-density polyethylene (same as in Fig. 17)

A similar example can be given also for high-density polyethylene used for manufacturing of oriented film threads [68,69]. Figure 17 illustrates the behavior of different brands of high-density polyethylene under longitudinal strain, and Fig. 18 gives the same under shear strain. Divergence of respective curves under shear strain is insignificant and at extension the time during which one and the same value of total strain is attained by different samples differs by a factor 15 (!).

Beside the molecular-weight distribution, the behavior of melts under longitudinal strain is affected by the ramification of macromolecules which can be seen from comparison of dependencies 2 and 4. The difference in the behavior of samples stems, apparently, from the fact that sample 2 contains 0.1 CH_3 groups per 100 atoms of carbon and sample 4 contains 0.24 CH_3 groups.

One more successfull practical application of the suggested method has been described recently [73] where the behavior of several brands of low-density polyethylene

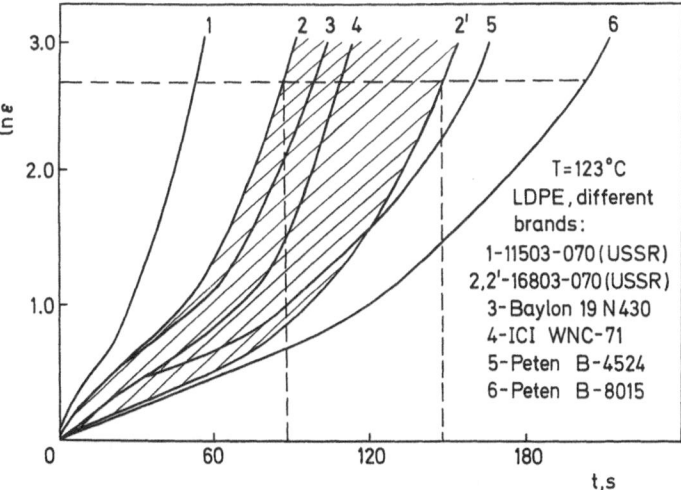

Fig 19. Variation of total strain (ln ε) with time (t) in extension under conditions of F = const for different brands of alkathene manufactured in different countries

are compared under longitudinal strain; the brands were those which are used for paper lamination. Processing of low-density polyethylene of brand 11 503-070 (USSR) was performed with great difficulties: the film canvas broke often, while the frequency of breaks was increasing with increase in the lamination process speed and reduction of temperature; however, the causes of failures in technology could not be identified with the help of characteristics of standard estimations of polymers. Determination of number-average and mass-average molecular weight (MM_n and MM_w), respectively, practically did not help to answer this question [73]. Flow curves, as in previous cases with polystyrene [72], practically coincided in different brands of low-density polyethylene manufactured in the USSR, on one hand, and Alkathene WNC-71, Bilon 19N430, Pethene B-8015, and B-4524, on the other hand. Differences between them could be identified only in the behavior of the melts of the above-mentioned samples of low-density polyethylene (see Fig. 19). It can be seen in Fig. 19 that low-density polyethylene of brand 11 503-070, at extension under constant-force conditions, reaches one and the same value of ε 1.5 times faster than all other brands of low-density polyethylene usually recommended by manufacturers for paper production.

Thus, tests of molten thermoplastics for extension at constant force turn out to be very informative and characteristic for estimation of 'technological adaptability' of raw materials. It should be pointed out that the suggested method of certification of thermoplastics with the help of tension tests at $F = const$ can be standardized by means of introducing a new index of technological properties suggested by the present authors. It is convenient to characterize the behavior of materials under longitudinal strain by the *melt extension index* (MEI) [69] which is actually the time during which the preset value of total strain (elongation) is attained in extension at preset constant force and at a certain temperature. Thus, in case of the method being standardized, it is necessary to establish only two parameters: the value of tensile force and temperature (similar to MFI where the load value and temperature of standard measurements are also preset). Prior to extension tests, the samples should be subjected to a simple processing, namely annealing, in order to exclude (or minimize) the effect of sample molding conditions upon the obtained results. The time of annealing is easy to be determined experimentally, usually it is between 10 and 30 min. but it should be determined, naturally, for each material in order to make sure that the results are independent of the conditions of sample manufacturing.

In view of the high sensitivity of the method to specific features of the polymer structure, it is convenient to use the MEI spread value to characterize the homogeneity of rheological and technological properties of the raw materials.

The suggested method of MEI identification is easily feasible under laboratory conditions, simple in terms of necessary equipment, easy for standardization which allows to recommend it both for research and commercial experience.

Let us consider briefly two more problems in the field of melt extension which are rather important in the thermoplastic polymer processing technology: identification of interconnections between longitudinal strain aWd structural characteristics of material and their alterations, as well as regularieties describing the process of melt drawing from jolding tool or the so-called 'jet drawing' (see also Sect. 4.3).

As discussed above, series of experiments and even a limited number of tests in longitudinal strain under constant-force conditions are 'representative' and very informative in terms of behavior of thermoplastics in processing. However, unfortu-

nately, these experiments were paid inadequate attention by researchers and, consequently, publications dealing with this matter are very few. That is strange, especially
taking into account that the first works which gave start, apparently, to the study of
extension of molten polymers were carried out by Kargin and Sogolova (exactly
under conditions of constant tensile force) as early as 1949, i.e., about 40 years ago [11,
12]. Experiments with polyisobutylene enabled the authors to propose division of
total strain into high-elastic (reversible) and residual (irreversible) components. The
high viscosity of the used PIBs (ca. 10^{10} Pa s) significantly simplified the procedure
of experiments since it eliminated difficulties associated with fixing of samples and
measuring their weight.

We should also mention an early work by Slonimsky and Askadsky [74] who were
apparently the first to observe structural changes taking place in extension under
conditions of constant force. Three characteristic sections (see Fig. 20) were identified
on the curves of strain versus tension time at $F = const$. These sections correspond
to polymer flow in the amorphous state, the process of molecular ordering and crystallization, and, finally, to polymer flow in the crystalline state. The presence of
crystalline formations on the latter section was detected with the help of X-ray-
structural and electron-microscopic investigation of extended samples. As the tensile
stress was lifted, the sample amorphised again and contracted. The occurrence of
a drastic increase in strain on the second section was accounted for [74] by exhaustion
of the longevity of supramolecular structures.

It should be noted that a radical growth of strain is observed always in extension
of the melts of both crystallizing and noncrystallizing polymers but, unfortunately,
structural investigations can not be carried out because of the high speed of the process
development at significantly lower viscosity of samples (e.g., polyolefins compared
to high-viscous polyisobutylene). Thus, the authors of Refs. [68–73] failed to observe
at least once the section of polymer flow (polyethylenes and even polyisobutylenes)
in the crystalline state which is associated, apparently, with the low molecular weight
of the materials compared, for example, to polyisobutylene, in which this had been
identified earlier [74].

It is interesting to consider the development of high-elastic (reversible) strain. It
grows abruptly on the first section, passes its maximum on the second section, and

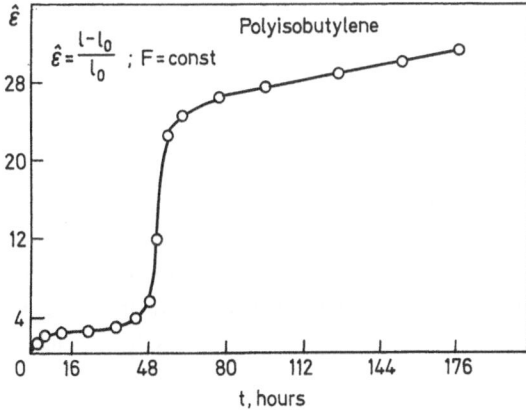

Fig. 20. Variation of total strain ε with time t in extension of polyisobutylene under conditions of $F = const$. [74]

on the third section it decreases gradually in the beginning and then practically does not change. A check of the known relationship:

$$\sigma = \eta_0(d\dot{\varepsilon}/dt) \tag{14}$$

where η_0 is initial viscosity, the constant of the material identically connected with the molecular weight of the polymer; $\dot{\varepsilon} = (1 - l_0)/l_0$ (here l_0 and l are the initial and current length of the sample, respectively) has revealed [68-71] that it describes the process satisfactory on the 1st and 3rd sections of the tension curve, but it is absolutely inadequate for description of the 2nd section when structural changes of the material take place.

Division of the total tensile strain under conditions of $F = \text{const}$ into several components [25,68,69] produced interesting results (see Fig. 8). It has been found that the behavior of molten low-density polyethylene (Fig. 8a) is qualitatively different from polyisobutylene (Fig. 8b) the extension of which was performed under temperature conditions where the high-elasticity modulus, relaxation time, and initial Newtonian viscosity practically coincided (in the linear range) in the compared polymers. Flow curves in the investigated range of strain velocities were also very close to one another (Fig. 21). It can be seen from the comparison of dependencies given in Fig. 8a,

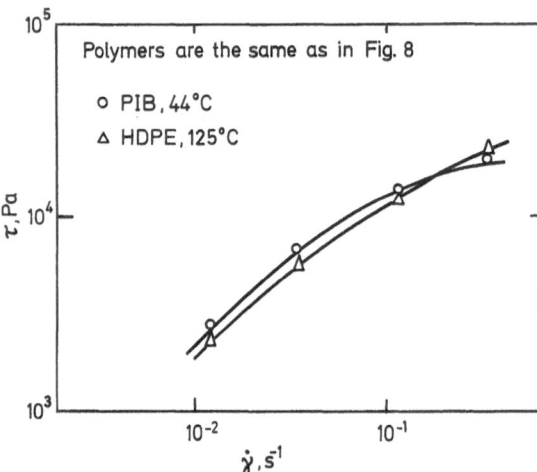

Fig. 21. Stress τ versus shear velocity $\dot{\gamma}$ (flow curves) for the same polymers and at the same temperatures as in Fig. 8

b that molten low-density polyethylene is extended 6 times slower than polyisobutylene. Strain velocity $\varkappa = d(\ln \varepsilon)/dt$ varies differently in extension of these polymers. In case of polyisobutylene the strain velocity increases continuously, and in case of low-density polyethylene it grows slightly, then reduces and then increases again. A practically similar character of the alteration of tensile strain velocity was observed also in the modified high-impact polystyrene [75]. Division of total strain into components has indicated that in case of extension of low-density polyethylene under constant-force conditions, there is a section in which there is practically no accumulation of irreversible strain, i.e., there is practically no flow and there is only an

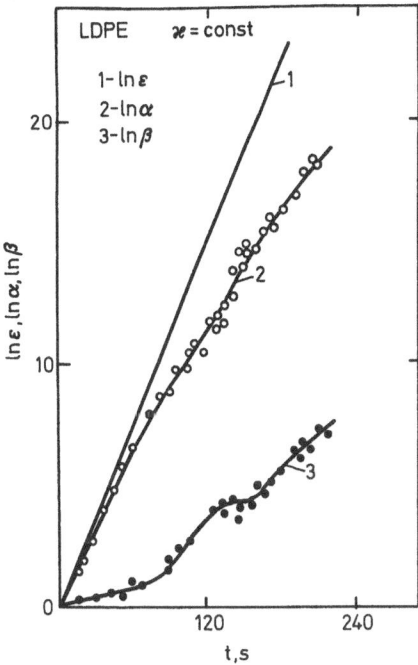

Fig. 22. Variation of total strain (ln ε) and its reversible (ln α) and irreversible (ln β) components with time (t) in extension under conditions of constant strain velocity (ϰ = const)

increase in the high-elastic reversible strain; the same effect was observed earlier [24] in the same low-density polyethylene under conditions of constant strain velocity, Fig. 22. Note that this type of behavior provides a possibility to accumulate a much greater reversible strain in the material at one and the same total strain. Therefore, the identification of the flow retardation effect which was already discussed in Sect. 2.2.7 is of great practical importance since it is well known (e.g. [2, 17, 19, 75]) that the strength of a polymer in the solid state is conditioned not by the total strain (drawing rate) but by the value of accumulated high-elastic strain correlating with the strengtening orientation effects.

A comparison was made between quantitative descriptions of dependencies of the total strain and stress upon time in extension under conditions of constant force with the use of several different models [76]: the known Maxwell's and Ostwald-De-Villes (power principle) models, Lodge model [50, 53, 77, 78], and Wagner model[79-82]. The best description (higher correspondence to experiment with polyethylene) was produced [76] with the help of the last of the above-mentioned models, and, as specified above, the model described in Refs. [46, 47]. However, it should be noted that in Ref. [76] the development of high-elastic (reversible) and irreversible components of the tensile strain at F = const were not treated separately which could be of significant interest.

In case of an accurate formulation of the problem, it is difficult to consider theoretically the process of molten thermoplastic drawing from a molding tool even under isothermic conditions. In the literature several approximate approaches have been suggested. For instance, it has been proposed [83, 84], for analysis of a similar problem (isothermic swelling of a jet leaving the capillary without subsequent drawing), to "cross-link" the inner (in the capillary) and the outer (free jet outside the capillary)

solutions by the flow of elastic energy. In this case the energy was calculated empirically and jet behavior at the outlet was regarded as purely elastic [84].

Another unidimensional approach [85] was to calculate the parameters of the ex-
tracted jet starting from the point on the central line of the jet corresponding to its
maximum swelling. This pattern is semiempirical and it needs a preset area of the
maximum cross section which is determined experimentally.

The inner (flow in the capillary) and the outer (outside the capillary) solutions of
the problem of polymer drawing from the molding tool were also "cross-linked" by
the flow of elastic energy [86] (the idea was borrowed from the above-mentioned
Refs. [84,85]) but this flow was calculated already accurately with the use of the model
given in Ref. [48], and the external extraction of the jet was calculated with regard to
the possible flow. Analytical data have been compared to the experiment to reveal a
satisfactory quantitative correspondence [86].

4.3 Rheological Analysis of the Processes of Film Molding from Melts

4.3.1 Sleeve Films

The rheology of the process of extrusion-inflation manufacturing of sleeve films is a
matter studied by many investigators. In a first approximation the behavior of molten
polymers, as viscoelastic fluids, under extension in the process of extrusion was regarded as similar to jet drawing of fibers (see, e.g., early works in this field by Han and
Park [87], Stivenson and Changue [88], etc.). However, in the process of extrusion with
inflation a tubular blank of a molten polymer is drawn in two directions: longitudinal
and lateral, therefore the process should be considered as a biaxial longitudinal flow
while the molding of fiber can be regarded as a uniaxial longitudinal flow.

The approach to analysis of biaxial extension of melts in the simulation of the sleeve
inflation process was developed by Pirson and Petrie in 1966–1970 with the use of
ideas of the thin shell theory which allows to substitute sleeve film by flat film in analysis. The problem was formulated more accurately and completely and solved in
works by Han et al. [89]. The author made several conclusion: the velocity of material
extension changes in the main direction of sleeve motion while effective longitudinal
viscosity may increase, decrease, or remain constant depending on the nature of material and the range of strain velocities under consideration; longitudinal viscosity
of the material at fixed process parameters decreases with temperature rise (the behavior of longitudinal velocity is described more strictly above, in Sect. 2.2.6).

It should be noted that Han's and Park's experiments were carried out under conditions of multiple biaxial extension and homogeneous biaxial extension in inflation
was studied experimentally by Denson et al. as early as 1971–1973.

In all above- and below-cited publications in this field (e.g. [84]) the problem was
solved in order to calculate the tensors of strain velocity and stress, to prognosticate
alteration of longitudinal viscosity, profile of alteration of the thickness of material
over the height of the film sleeve (by coordinate on the central line of the sleeve counted
from the outlet face of the extrusion head) and configuration of the sleeve ("bubble")
and also to solve thermal problems in order to determine the dependency of melt
temperature upon height (or time) and to forecast the position of the crystallization

line (or hardening) of the polymer. Solutions of the above-mentioned problems can be found in the above-cited publications and in review [7], therefore there is no need to discuss them here.

The interest for melt extension in molding of sleeve films has not become less during recent years. A number of authors made attempts to improve analysis and modelling of the process in view of the following major problems: 1) analysis and elucidation of strain prehistory and its effect upon the state of a melt leaving the circular heads of different design [90] (the final results here are analytical dependencies used to determine the relationship between normal stresses and tangential stresses depending on geometrical characteristics of heads which is recommended as criteria for selection of optimum design of the molding tool); 2) analysis of the sleeve cooling process in order to improve the accuracy of prognostication of its profile and to develop recommendations on intensification of the process [91,92]; 3) simulation and analysis of the process with regard to specific behavior in extension of a given molten material (primarily when manufacturing of films made of new raw materials is started, for example, copolymer of ethylene with propylene [93]); 4) development of continuous quality control methods (to control thickness which changes under extension, orientation effects under drawing, etc.) [94]; 5) investigation into the stability of the film sleeve molding process in plants of different design and under different process conditions [95]; 6) development of automatic microprocessor-controlled production lines (which calls for analytical models adequate for selection of controlled process parameters and process control elements) [96-98].

Note that problems 4–6 of the above-mentioned groups of research guidelines have become dominating during recent years.

4.3.2 Flat Film Canvas

Molding of film canvas (blank) is one of the most complicated stages in the technology of flat film manufacturing from molten polymers. Actually, jet drawing could be also simulated in this case in a first approximation like it is done for fiber molding by unidimensional (uniaxial) extension. However, this rough approximation does not reflect many real and significant factors seriously affecting the process and therefore it has no practical value [7].

Molding of flat film canvas is usually performed by pouring a melt coming out of a flat-slit extruder head (spinneret) on the receiving drum (a roll with sufficiently

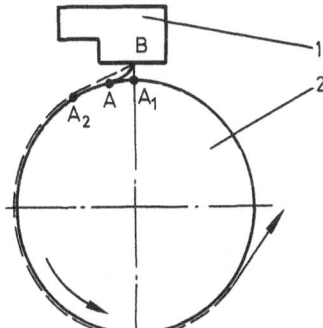

Fig 23. Diagram of molding of molten polymer flat film canvas: 1 — flat-slit extrusion head (spinneret); 2 — receiving roller ("drum")

large diameter) rotating at a constant speed (Fig. 23); jet drawing in the longitudinal direction takes place in the clearance between the head and the drum due to the fact that the linear velocity of melt extraction exceeds the velocity at which it is fed by the extruder via the spinneret. In spinneret drawing a polymer is most susceptible to various disturbances, so this stage determines primarily the quality (first of all thickness variations, i.e., uniformity of thickness) of the film canvas. A polymer contacting with the surface of the receiving-cooling drum sticks thereto, the process of extension ceases, the material is cooled down and removed from the drum in form of continuous canvas. The latter can serve either as a finished product (comparatively thick film, sheet, or strip) or as a blank for further orientation uni- or biaxial drawing of polymer in the high-elastic state in manufacturing of thin (10–100 mm) and ultrathin (1–10 mm) oriented films.

Usually a melt is molded vertically downwards, however, as it is seen in Fig. 23, the molded canvas is located in the extension zone at an angle (which sometimes may be rather large) to the vertical. Canvas in the extension zone can be practically flat (dashed line in Fig. 23) or may sag under the action of gravity (continuous line in the same figure) which is determined by the specific molding conditions, primarily by the tensile force and thickness of the extruded blank.

An attempt to study biaxial extension of flat molten polymer canvas was made [99]. The major achievement of this work is, apparently, the development of experimental plant and research procedures.

Jet drawing of flat film canvas in real technologies (not in model studies of general problems concerning drawing of a jet from a round capillary) has not been discussed adequately in the literature. A special case is the known publication by Mzelsky [100] dealing with this problem. The results of this work found a practical application in form of a number of useful recommendations on optimization of canvas-blank molding in manufacturing of thin oriented films of polystyrene, polypropylene, and polyethylene terephthalate.

Reference [99] focuses on qualitative and approximate quantitative analysis of the occurrence of inhomogeneities in film canvas in longitudinal and lateral directions The inhomogeneity of blank film in the longitudinal direction [99] is caused by disturbances of different nature while that in lateral direction is a consequence of canvas drawing being performed not under conditions of simple tension. Strained sample inhomogeneity can be illustrated by diagrams given in Fig. 24. The square element in areas at the edges of the blank under tension tends to take the form of a rhombus (in Fig. 24 it is indicated in the middle of the film for reference purposes) and forces occur (stresses) causing lateral contraction of the sample. On the hole, the square grid marked on the blank sample is distorted insignificantly in the central part of the canvas or remains unchanged (this part of the canvas is subject to simple shear) while the edges of the sample are curved significantly and are under conditions close to simple tension.

There are also other effects accompanying the extension of molten polymer in the form of canvas in manufacturing of flat films which are well known from technological experience: firstly, the thickening of the edges of the sample compared to its middle and formation of "rims"; secondly, the lateral contraction does not spread over the full width but covers only the edges of the canvas. This is, in fact, a boundary effect disturbing the homogeneity of material orientation. As length increases, the extended

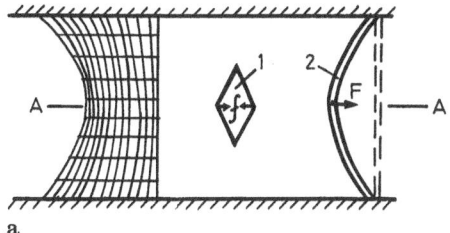

a

A—A

b

Fig. 24a, b. Film sample after extension: **a** view of square grid marked on the sample prior to extension (on the extended sample); **b** section A—A of the sample given in **a**. View of the square (1) and belt (2) elements of the film after extension

boundary area covers the larger section of the film (i.e., approaches the central line of extrusion and the longitudinal central line of the canvas). This is, in principle, the background of an important technological recommendation: to improve film homogeneity in terms of property and thickness, the jet-drawing zone (the distance between flat-slit extrusion head and receiving cooling shaft, the drum) should be shortened and the width of the canvas should be increased.

An interesting point is also the qualitative analysis and quantitative estimations given for disturbances causing the inhomogeneity of film thickness in its longitudinal direction [100]. Figure 25 gives a wedge-type lateral cross section of a melt in the jet drawing zone (boundary area). At constant longitudinal viscosity, the generator of the wedge takes the form of an exponent of the following type:

$$H = H_0 \exp\left[-F/(\lambda Q) \cdot x\right] \tag{15}$$

where H_0 and H are the initial and the current (at longitudinal coordinate X) thickness of molten polymer film; $F = const$ is the force of drawing Q the flow of polymer (process output) $\lambda = const$ the Trutonian viscosity of the melt.

An attempt was made to assess quantitatively the profile of the so-called curvilinear

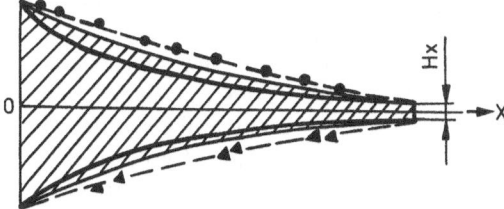

Fig 25. Profile of molten polymer film canvas in the jet drawing zone ("extension wedge"), the exponential wedge is section-lined. Continuous line corresponds to wedge profile at longitudinal viscosity increasing in the process of extension; dashed line corresponds that at decreasing viscosity. Symbols ▲ indicate experimental data obtained in extension of flat polystyrene threads; ● indicates data obtained from (21)–(22) [100] at $\tau = 1$; $a = 0.98$; $V_0 = 1$; $P_m/(Q_\eta) = 0.882$; $H_0 = 1$

extension wedge for real cases when the longitudinal viscosity of the melt does not remain constant [100] (see Sect. 2.6.6). In this case, to simplify calculations, a rough approximation of function $\lambda(\dot{\varepsilon})$ by an easily integrable expression of hyperbolic type: was used:

$$\lambda = A \pm (K/\dot{\varepsilon}) \tag{16}$$

where the "minus" sign belongs to the increasing and the "plus" sign to the decreasing viscosity of the melt under extension at a strain velocity.

Certainly, Eq. (16) is only a formalized representation of the function $\lambda(\dot{\varepsilon})$ representing in a simplified form the non-constancy of viscosity (see Sect. 2.2.6 for details); however, it provides a possibility to assess the tendency of extension wedge form variation. According to estimations [100], at $\lambda \neq$ const, the wedge deviates from the exponential one, it becomes more sharp for λ increasing with $\dot{\varepsilon}$ and more smooth for decreasing viscosity (see Fig. 25).

In estimation of the stability of the jet drawing process [99] a so-called step-wise disturbance of the canvas ($\Delta H_0/H_0$) was proposed at the inlet to the extension zone and its variation ("germination") during drawing. The associated major conclusions correspond, in principle, to observations of engineers in manufacturing of oriented thermoplastic films. These conslusions include the following:

In case of $\lambda =$ const, the extension of canvas by α times results in an increase in disturbance by the same factor:

$$\Delta H/H = \alpha(\Delta H_0/H_0) \tag{17}$$

In case of $\lambda(\dot{\varepsilon})$ being an increasing function, the primary disturbance "germinates" slower than it follows from Eq. (17); the higher the drawing ratio α, the slower it grows. In case of a decreasing function $\lambda(\dot{\varepsilon})$ the situation is quite opposite: the higher the rate of decrease in longitudinal viscosity at increasing $\dot{\varepsilon}$, the faster the disturbance "germinates".

Analysis of the melt molding process usually takes into account only the effect of tensile forces upon the material [9, 87-89]. However, in the general case forces of other nature can be also significant: gravitational, inertial, surface tension, and friction forces. Engineering estimations of the influence of these forces upon the profile of the canvas cross section under drawing and stability of the molding process have been given [100]. These simplified calculations have been performed with several assumptions: melt canvas is fed vertically to the receiving-cooling drum, melt swelling at outlet of the molding heas is neglected, melt temperature in the short zone of jet drawing does not change, and the melt flow is assumed to be of Newtonian character.

The latter assumption concerning the absence of anomalous viscosity in molten polyethylene terephthalate (PETP), which was taken as an example for specific calculations, is quite realistic, especially for low-viscous film brands of polyethylene terephthalate as it is demonstrated in a detailed study of rheological properties of this thermoplastic [101]. Conclusions made in Ref. [100] are the following: the gravitational force can be commeasurable with the drawing force while the friction force is insignificant and the inertial and surface forces also do not play any notable role (in the specific example they account for 3% and 8% of the tensile force, respectively). It should be, however, noted that the above relationship of forces may change under varying conditions of molding. Thus, for example, the rate of the process being increased by factor of 3 results in the increase in tensile force and surface tension forces also by a factor of 3, in this case the fravitational component does not change and the inertial one increases by a factor of 9; in other words, the role of the fravitational component will be reduced, that of the surface component will not change, and the role of the inertial component will increase. Not that the effect of the above forces becomes significant only due to the relatively low viscosity of molten polyethylene terephthalate [100, 101] and in case of other thermoplastics, the viscosity of which is higher by 1.5 to 2 orders of magnitude, their contribution can be well neglected.

The uniformity of thickness of the manufactured film is greatly affected by pulsations of extrusion velocities (melt flow) and the rate of melt extraction by the receiving-cooling drum. The causes of flow pulsations at the outlet of the extruder and measures to reduce them have been analyzed [102]. As regards the effect of drum's speed pulsations upon the non-uniformity of the blank film, the latter is caused not only by variations of the drawing force resultant from that effect, but also by the displacement of the line of contact between the molten polymer and the cooling surface of the drum. In real processes the melt canvas is directed to the drum not vertically downwards but at a certain angle to the vertical and sticks to the surface of the cooling element. Even at a short-time drop of the drum's speed the canvas sags and extension of material practically ceases in this section which results in a significant non-uniformity of the thickness of manufactured film in the longitudinal direction [100].

Certainly, it is impossible to discuss all significant applications of the theoretical and experimental study of molten polymer extension in one review (even in an extensive one). In the last section we treated only several technological problems of current concern the solution of which is based on the regularities of extension of molten thermoplastic. The selection of problems stems not only from potential advantages which may be provided by the use of information accumulated in the studies of melt extension but, naturally, also from personal scientific and technical interests of the authors. Specialists will easily extend this brief review of practical applications of the knowledge on molten polymer extension with examples familiar to them from their scientific and practical experience. The authors considered the objective of the final section of this publication to illustrate the technological significance of melt extension by clear examples. The scientific value of this field of rheological studies is generally recognized nowadays.

The analysis indicates that presently quite adequate phenomenological models are available for description of the straining of commercial (polydisperse) polymers in the liquid state. A comparatively clear understanding of the mechanics of the processes of manufacturing of sleeve-type and flat films of molten thermoplastics also has been developed. So far, physical approaches have provided rheological models only for monodisperse polymers (the properties of which differ significantly from those of the ones used in industry).

Apparently, in the nearest years we may expect new results in the field of phenomenological models which will become closer to molecular characteristics of polymers. This will facilitate a directed control of the properties of finite productes effectuated by means of variation of the molecular characteristics. At the present stage the above-discussed method of certification of raw materials proposed by the authors of this review and AN Prokunin may be useful for molding process control.

5 References

1. Cassale A, Porter RS (1979) Polymer stress reactions. Academic Press, New York, San Francisco, London; (1983) Chimia, Moscow
2. Fridman ML (1987) Study and assessment of technological properties of polymers. In: Wolfson SA (1987) Fundamentals of the development of polymer production technologies. Chimia, Moscow, pp 185—231
3. Vinogradov GV, Malkin AYa (1980) Rheology of polymers. Mir Publishers, Moscow
4. Malkin AYa (1985) Rheology in polymer technology. Znanie, Moscow
5. Vinogradov GV, Malkin AYa, Volosevitch VV (1975) J. Appl. Polym Sci 22: 47
6. Everige AE Jr, Ballman RL (1976) J Appl Polym Sci, 23: 1137
7. Fridman ML (1977) Crystalline polyolefins processing technology. Chimia, Moscow
8. Fedorovskaya TS (1988) Himicheskaya Promyshl. za Rubezhom 1: 32
9. Petrie CIS (1979) Elongational flows. Pitman, London, San Francisco, Melbourne
10. Takahashi M, Masuda T, Onogi S (1983) J Soc Rheol Japan 11: 13
11. Kargin VA, Sogolova TI (1949) Zhurn Phys. Chimiy 23: 540
12. Kargin VA, Sogolova TI (1949) Zhurn Phys Chimiy 23: 550
13. Leonov AI, Prokunin AN, Vinogradov GV (1970) In: Vinogradov GV (ed) Achievements of polymer rheology, Chimia, Moscow, pp 41—51
14. Vinogradov GV, Leonov AT, Prokunin AN (1969) Rheol Acta 8: 482
15. Nitshman H, Schrade J (1948) Helv Chim Acta 31: 297
16. Ballman RL (1965) Rheol Acta 4: 137
17. Meissner J (1969) Rheol Acta 8: 78
18. Messner J (1971) Rheol Acta 1: 230
19. Radushkevitch BV, Fichman VD, Vinogradov GV (1970) In: Vinogradov GV (ed.) Achievements of polymer rheology. Chimia, Moscow, pp 24–39
20. Fichman VD, Radushkevitch BV, Vinogradov GV (1970) In: Vinogradov GV (ed) Achievements of polymer rheology. Chimia, Moscow, pp 9–24
21. Munstedt H, Laun HM (1979) Rheol Acta 18: 492
22. Cogswell FN (1969) Rheol Acta 8: 187
23. Prokunin AN (1978) Nonlinear elastic effects in extension of polymer fluids. Experiment and Theory. Preprint. Inst. Appl Mech Acad Sci USSR, Moscow, No 104
24. Prokunin AN, Filippova NP (1979) Inzhenerno-Physichesky Zhurnal 37: 724
25. Prokunin AN, Sevruk VD (1981) Inzhenerno-Physichesky Zhurnal 41: 74
26. Prokunin AN, Proskurnina NG (1979) Inzhenerno-Physichesky Zhurnal 36: 42
27. Fichman VD, Radushkevitch BV, Goldina EG, Vinogradov GV (1974) Mehanika Polymerov 1: 137
28. Cogswell FN (1972) Trans Soc Rheol 16: 383
29. Laun HM, Munstedt H (1976) Rheol Acta 15: 517
30. Laun HM, Munstedt H (1978) Rheol Acta 17: 415
31. Stevenson YE (1972) AIChE Journ 18: 540
32. Ishikura O, Koyamo K (1977) Polymer 21: 164
33. Matsumoto T, Bogue DC (1977) Trans Soc Rheol 21: 453
34. Akutin MS, Prokunin AN, Proskurnina NG, Sabsay OYu (1977) Mehanika Polymerov 2: 353
35. Prokunin AN, Proskurnina NG (1979) Inzhenerno-Physichesky Zhurnal 36: 504
36. Sabsay OYu, Koltunov MA, Vinogradov GV (1972) Mehanika Polymerov 4: 750
37. Sevruk VD, Prokunin AN (1980) Manifestations of flow retardation effect in extension of molten polyethylene at constant force. 2nd All-Union Symposium: Theory of mechanical processing of polymer materials. Perm, pp 165–166
38. Slonimsky GA, Musaelyan IN (1964) Vyskomolecularniye Soedineniya 6: 101
39. Leonov AI, Prokunin AN (1983) Rheol Acta 22: 137
40. Prokunin AN (1980) J Polym Mater 8: 303
41. Prokunin AN, Sevruk VD (1982) Inzhenerno-Physichesky Zhurnal 42: 987
42. Upadhyay RK, Isayev AI (1984) J of Rheol 28: 581
43. Upadhyay RK, Isayev AI, Shen SF (1963) J of Rheol 27: 155
44. Upadhyay RK, Isayev AI, Shen SF (1981) Rheol Acta 20: 443

45. Leonov AI, Lipkina EH, Pashkin ED, Prokunin AN (1976) Rheol Acta 15: 411
46. Prokunin AN (1988) Xth International congress on rheology Sydney, Australia, August
47. Prokunin AN (1988) Inzhenerno-Physichesky Zhurnal 54: 221; (1988) Rheol Acta
48. Leonov AI (1976) Rheol Acta 15: 85; Leonov AI, Prokunin AN (1980) 19: 393
49. Green MS, Tobolsky AV (1946) J Chem Phys 14: 80
50. Lodge AS (1956) Trans Faraday Soc. 52: 120
51. Yamamoto M (1956) J Phys Soc Japan 11: 413
52. Doi M, Edvards SF (1979) J Chem Soc Faraday Trans 75: 38
53. Lodge AS (1964) Elastic liquids. Academic Press, New York, London
54. Volkov VS, Vinogradov GV (1984) Rheol Acta 23: 231
55. Curtis CF, Bird BB (1981) J Chem Phys 74: 2016
56. Volkov VS, Vinogradov GV (1985) J Non-Newtonian Fluid Mech 18: 163
57. Volkov VS, Vinogradov GV (1987) 25: 261
58. Volkov VS, Vinogradov GV (1988) Progress and trends in rheology II. In: Giesekus H, Hibberd MF (ed) Steinkopff Verlag, Darmstadt
59. Long JH, Muller R, Frolich D (1986) Polymer 27: 6
60. Volkov VS (1984) Intern. Rubber Conference, Moscow, Preprint A67
61. Wang RH (1963) Model and Simul, vol 14. Proc. 14th Annu. Pittsburg Conf 21–22
62. Kalinchev EL, Sakovtzeva MB (1983) Properties and processing of thermoplastics, Chimia, Leningrad
63. Fridman ML, Malkin AYa (1976) Plasticheskiye Massy 8: 23
64. Driscoll SB (1980) Rubber World 3: 31
65. Sipdzi X (1983) Japan Patent 58-119844
66. Ulyanov LP, Sabsay OYu, Fridman ML et al (1988) Author's Certificate No. 1377662 (USSR)
67. Winter HH (1983) Pure and Appl Chem 55: 943
68. Sevruk VD (1984) Extension of molten thermoplastics at outlet from molding tool. Thesis. Inst Fine Chem Tech, Moscow
69. Sevruk VD, Prokunin AN, Fridman ML (1984) Regularities of molten polymer extension and their manifestation in plastic processing. NIITEHIM, Moscow
70. Prokunin AN, Sevruk VD (1980) Inzhenerno-Physichesky Zhurnal 39: 343
71. Sevruk VD, Prokunin AN, Fridman ML (1980) Regularites in extension of molten thermo-plastics. In: Fridman ML (ed) Rheology in polymer processing. NPO Plastic, Moscow, pp 84–99
72. Sevruk VD, Prokunin AN, Fridman ML, Novikov DD (1984) Plasticheskiye Massy 7: 61
73. Sevruk VD, Blinova NK, Kalashnikova OD (1988) Plasticheskiye Massy 2: 22
74. Slonimsky GL, Askadsky AA (1967) Mehanika Polymerov 4: 659
75. Chalaya NM, Sabsay OYu, Vinogradov GV et al. (1982) Specific rheological and technological properties of modified polystyrene. In: Fridman ML (ed) Processing of filled compound materials. NPO Plastic, Moscow, pp 80–90
76. Raible T, Stephenson SE, Meissner J, Wagner MN (1982) J Non-Newtonian Fluid Mechan 11: 239
77. Lodge AS (1968) Rheol Acta 7: 379
78. Lodge AS (1974) Body tensor fields in continuum Mechanics. Academic Press, London, New York
79. Wagner MH (1976) Rheol Acta 15: 136
80. Wagner MH (1979) Rheol Acta 18: 83
81. Wagner MH, Raible T, Meissner J (1979) Rheol Acta 18: 427
82. Wagner MH, Meissner J (1980) Macromol Chem 181: 1533
83. Grossley WW, Glasscock SD, Crawley RL (1970) Trans Soc Rheol. 14: 519
84. Malkin AYa, Goncharenko VV, Malinovsky VV (1976) Mehanika Polymerov 3: 487
85. Denn MM, Petrie CJS, Avenas P (1975) AIChE Journal 21: 791
86. Leonov AI, Prokunin AN (1984) Rheol Acta 23: 62
87. Han CD, Park YY (1975) J Appl Polym Sci 19: 3257
88. Stevenson JP, Chung SCK (1974) Paper presented at Ann Meeting Soc Rheol 45th Amherst, Mass, pp 21–24
89. Han CD (1976) Rheology in polymer processing. Academic Press, New York, San Francisco, London; (1979) Vinogradov GV, Fridman ML (eds) Chimia, Moscow
90. Winter HH, Fischer E (1981) Polymer Engng and Sci 21: 366

91. Michaeli W, Menges G (1978) 37th Ann Techn Conf Soc Plast Engng New Orleans, La, pp 141–145
92. Lohse G, Marinow S (1986) Plaste und Kautschuk 33: 106
93. Speranskaya TA, Goldin PO, Kreizer TV et al. (1982) In: Modelling and equipment of plastic manufacturing processes. Leningrad, pp 18–27
94. Menges G, Winkel E, Nordmeier J (184) Papier und Kunststoff-Verarb 19: 44
95. Minoshima W, White J (1983) Polym Engng Rev 2: 212
96. Malik K, Lev V, Matousek Z (1981) Instrum and Automat: Paper Rubber Plast and Polym Ind Proc 4th IFAC Conf. Chent 3–5, June 1980. Oxford pp 103–108, 458–460
97. Breier J, Kathe H, Marx D, Dorsch HT (1981) Plaste und Kautschuk 28: 217
98. Grigoriasi V, Petrovan S (1985) Mater Plast 22: 193
99. Meissner J, Stephenson SE, Demarmels A, Portman P (1982) J Non-Newtonian Fluid Mechanics 7: 10
100. Mzelsky AI (1980) Rheological analysis of the process of film canvas molding from molten polymer. In: Fridman ML (ed) Rheology in polymer processing. NPO Plastic, Moscow, pp 71–84
101. Dubinsky MB, Sabsay O Yu, Fridman ML, Mzelsky AI (1986) Plasticheckiye Massy 3: 20
102. Fridman ML, Mikhailov SN, Muhametgaleyev AM (1988) Mathematical modelling of single-screw extruders. ZINTIHimnneftemasch, Moscow

Molding of Polymers under Conditions of Vibration Effects

M. L. Fridman and S. L. Peshkovsky
USSR Research Institute of Plastic Materials,
Perovsky proezd 35, Moscow, USSR

This paper reviews the results of investigations into low-frequency mechanical and high-frequency (ultrasonic) vibration effects upon flowable polymeric systems, primarily, on molten commercial thermoplastics. We tried to systematize possible techniques to realize vibration in molding of polymers. Theoretical and experimental corroboration is provided for major effects obtained at cyclic (shear and bulk) strains of molten polymers and compositions based thereon. It is demonstrated that combined stress of polymeric media is attained under overlapping vibrations and this results in a decreased effective viscosity of the melts, a drop i the pressure required to extrude them through molding tools, increased critical velocities of unstable flow occurrence and a reduced load on the thrust elements of extruder screws.

Power consumption aspects have been critically analyzed to reveal that vibration effects in molding heads may reduce specific power consumption in extrusion machines, while simultaneously improving the quality of manufactured products.

The paper gives an overview of effects occurring or acoustic treatment of dissolved and molten polymers. Emphasis is made on acoustic cavitation discovered recently not only in low-viscous fluids but also in molten polymers. Major guidelines have been specified for practical utilization of acoustic treatment of flowable polymers in molding: intensification of extrusion processes, reduction in thickness of produced films, directed mechanical destruction, chemical "activation" of melts, etc. Efficiency of overlapping high-frequency vibrations in molding of molten thermoplastics is discussed in terms of power consumption.

Advances in Polymer Science 93
© Springer-Verlag Berlin Heidelberg 1990

1 Introduction

Progress in the field of processing of polymer materials into various products has been rather extensive until recently: the diameter and length of extruder working elements were increased as did the, maximum injection volume of molding machines (automatic thermoplastic machines), speed of screws, etc. Machines have been developed to granulate polymeric materials at a rate of 10–20 t/h, to produce films with a width of 24 m and more, to mold products weighing 10–20 kg or even more than that, to perform oriented extrusion of ultrathin (1–5 mcm) large-size (up to 4 m) films, to extrude sheets and plates with a thickness of 20 mm and above at a width of 2–3.5 m. The attention of science and technology is focused on the major tendencies in the progress of technology and equipment intended for molding of semifinished items and plastic products which are discussed regularly in articles and reviews published in different countries including the USSR (see, for example, [1–5]). The achievements made witherto are impressive, however, this line of progress approaches the ultimate limit of its potential and, apparently, a further increase in dimensions (and, consequently, power and metal consumption) of processing machines cannot be regarded as a major area of development in this field.

This brought about a keen interest in other methods of intensification in processing. Lately, the directed effect of physical (mechanical) fields on molten polymers has become one such area. These effects, as demonstrated in many works published in the 1970s and in the 1980s, (see for examples [6–9]) result in altered parameters of micro- and macrostress of the system. Molding under conditions of directed physical fields, in particular, in the case of mechanical and acoustic vibration effects upon melts, is performed so that an additional stress superimposed on the polymer's main shear flow and the state of material is characterized by combined stress.

In the early works by Soviet scientists (primarily the followers of P. A. Rebinder and G. V. Vinogradov) and Western investigators (H. C. Booij, R. I. Tanner, J. M. Simmons, T. Kataoka, R. Osaki, et al.) published between 1966 and 1968, the vibration was proven to produce a powerful effect on rheological properties of two-phase disperse systems, filled polymers, rubbers as wellas dissolved and molten polymers.

We do not intend to discuss aspects of the use of cyclic strain conditions to obtain information on viscous, visco-elastic, and relaxation properties of polymers, although this, by itself, may present a significant interest for theoretical and experimental research. The aim of this publication is to review recent works which provide a basis for various technical applications, i.e., facilitate rearrangement of molding processes.

In principle, the combined stress in polymer fluid can be attained, for example, with the help of rotation longitudinal or lateral (in relation to the central line of the extrusion flow)-vibrations of the elements of molding tools. The systems of this type, characterized by comparatively low frequencies and significant amplitudes of effects, can be classified tentatively as mechanical systems, the working element of which has circular channels (formed by the casing (tip) of the extrusion head and the core), with one or two movable walls. Another group of effects characterized, on the contrary, by high frequencies but very small amplitudes, includes shear and bulk acoustic effects of ultrasonic frequencies (most often the commercial frequency of standard ultrasonic generators is about 20 kHz). In generalized form both above-mentioned

groups of effects of physical fields upon molten polymers provide a basis for technology which has been classified in the recent years as a separate group of processing
methods: vibromolding of polymers [10]. Vibration at infrasonic, sonic, or ultrasonic
frequency permits, at comparatively low power of the additional drive, the imparting
of very significant variable-sign (cyclic) velocities and accelerations to revolving or
vibrating elements and the particles of processed material which cone intro costact
with then. Due to this, a polymer is subject not only to the combined shear strain,
but also to a number of physical and chemical phenomena which intensify the processing, thus reducing power consumption of equipment and improve the quality
of finished products.

2 Mechanical Vibration Effects upon Molten Thermoplastic Polymers

1.1 Brief History of the Spiral Flow Theory

The simplest case of a combined shear is the spiral, or helical, flow occurring when a
material is extruded through a circular slit under the action of continuous pressure
difference while the cylinders are rotating with constant frequencies (or one of them,
for example, tip or core of the extrusion head. The concept of spiral (helicyl) flow was
introduced and the general problem was formulated by S. Rivlin in 1956. His work,
as well as a number of subsequent works, was focused only on the stable flow and
revealed, in general form, correlations for stress tensor components. Experimental
investigations into the character of the flow of a material with pronounced non-
Newtonian properties (for example, disperse systems — viscoplastic lubricants)
were carried out by G. V. Vinogradov, A. A. Mamakov, V. I. Pavlov and N. V. Tyabin from 1959–1960, with special care given to the increased-flow effect on rotation
of one of the molding elements. Spiral flow was compared to axial or concentric flows
in the narrow clearances between two aligned cylinders within a sufficiently wide
range of strain velocities. Strain velocities were determined under conditions of insignificant effect of the circular and axial flows upon one another. It has been revealed
that with increasing pressure difference (the cylinder speed of rotation being Ω = const),
in the flow area of a disperse viscoplastic system adjacent to the centre of the circular
clearance, the apparent ("effective") viscosity grows, a "nucleus" of the flow is formed
and flow conditions approach the "plug" type.

In 1957 P. Pasley and A. Slibar produced a solution of the particular problem of
spiral flow for Bingham-Shvedov fluid in the case of a minor contribution of rotation
to the distribution of total stresses in the circular clearance compared to axial flow.
In 1961 Myasnikov V. P. [11] and A. M. Gutkin [12] solved the problem for Bingham-
Shvedov body, having lifted a number of limits set forth by Pasley and and Slibar. A
diagram of states has been constructed as the result of general analysis of flow equations. The diagram reflected the qualitative character of the profiles of longitudinal
and angular velocities of pressure flow of Bingham-Shvedov fluid in a circular channel
at rotation of the inner pipe.

Numerical calculations of helical flow in non-linear viscoplastic fluid have been carried out comparatively recently by E. P. Schulman and A. N. Prokunin et al. [13, 14]. They have demonstrated the influence of flow and speed of rotation on the value of pressure losses and moment on the inner pipe dependent on the values of rheological parameters of the fluid.

The results of the latest research into helical flow of viscoplastic fluids (media characterized by ultimate stress or "yield point") have been systematized and reported most comprehensively in a recent preprint by Z. P. Schulman, V. N. Zadvornyh, A. I. Litvinov [15]. The authors have obtained a closed system of equations independent of a specific type of rheological model of the viscoplastic medium. The equations are represented in a criterion form and permit the calculation of the required characteristics of the helical flow of a specific fluid. For example, calculations have been performed with respect to generalized Schulman's model [16] which represents adequately the behavior of various paint compoditions, drilling fluids, pulps, food masses, cement and clay suspensions and a number of other non-Newtonian media characterized by both pseudoplastic and dilatant properties.

As regards the general case of viscoelastic fluids (including the thermoplastics characterized by pseudoplastic properties), the solution of the helical flow problem stems from the classical works by B. Colleman, W. Noll, H. Markwitz and A. Fredrickson carried out between 1959 and 1966. An important result of those works was the conclusion that combined shear is a viscosimetric flow and the knowledge of three viscosimetric functions determined from simple shear tests is vital for the complete solution of the problem. B. Colleman and A. Fredrickson used practically one and the same approach to obtain general relationships for the distribution of longitudinal and angular velocities in a channel and for the flow of a generalized non-Newtonian fluid [17, 18]. A. Dirkes and W. Showalter used the Oldroyd model to obtain similar non-linear relationships and compared the theory to the experiment carried out with a polymer solution for only one combination of angular velocity and average axial velocity of the flow. The obtained data corresponded well to calculations by B. Colleman [17] and A. Fredrickson [18]. R. Tanner in [19] discussed as analytical solution to the problem for a case of pseudoplastic polymer flow in a flat-slit head, and I. Savins and G. Wallick [20] considered a solution for a concentric head. They proposed a procedure for the numerical solution to the problem and qualitative assessment of viscosity alteration depending on the width of the circular slit. They have demonstrated that the viscosity of a molten thermoplastic grows at certain combinations of the values of angular velocity Ω and acial flow velocity, which had been observed earlier by G. V. Vinogradov et al. in experiments with disperse systems (lubricants).

Flow through a slit head has been studied also in polymers using the power-function flow principle (Ostwald-de Villes model). Some of the first works in this field were carried out in 1967–1968 by N. V. Tyabin et al. [21, 22]. Tests were arranged with molten polyethylene in an extrusion head having a rotary conic core. The divergence between theory and experiment was comparatively significant which, apparently, is accounted for, primarily, by the basence of a necessary similarity between the conducted experiment and the theoretical description of the flow in a flat slit.

A. N. Prokunin and M. L. Fridman [23, 24] seem to be the first investigators to have done the first adequately accurate calculations of power-function fluid helical flow in a circular (tubular) head with a rotary core. At the same time a series of experiments was carried out on laboratory and pilot plants with molten polypropylene of different brands and a satisfactory coincidence between the experimental and analytical data was attained.

It has been demonstrated in a number of the above-cited works (by S. Rivlin, G. V. Vinogradov et al., I. Savins and G. Wallick, N. V. Tyabin et al., A. N. Prokunin & M. L. Fridman) that the average velocity of the axial flow grows, the pressure difference being constant, with as increase in the value Ω within certain limits of the curve's section where viscosity strictly decreases.

The dependency of melt properties upon temperature has not been taken into account in all the above-mentioned publications. An attempt to calculate the temperature field and the rate of establishment of equilibrium temperature profile determined by thermal dispersion and thermal conductivity was made by H. Winter[25, 26].

A publication by B. Colleman, H. Markwitz and W. Noll[27] describes in detail the theory of viscosimetric flows which was further used by a number of investigators (see, for example, [23, 24, 28, 29]) to analyze axial flow in a clearance between motionless and rotating cylinders. The authors of [27, 30] have demonstrated theoretically that the spiral flow can be considered as a mutual overlapping (superimposing) of two strains occurring in a simple shear (see Sect. 1.2).

To conclude this brief digression into history, we may point out one more important aspect: the high efficiency of the combined shear in molding of filled thermoplastics. One of the first works in this field was [31] which described experiments carried out with polypropylene filled with a disperse aggregate calcium carbonate (chalk) and a short-fiber material-asbestos.

2.2 The Theory of Spiral (Helical) Flow of Molten Thermoplastic Polymers

The theory of viscosimetric flows, as mentioned above, permits the consideration of a spiral flow (Fig. 1) as a superimposition of two simple-shear strains. This approach

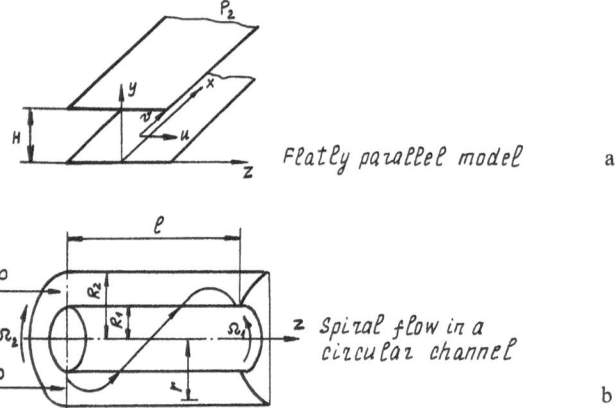

Fig. 1a, b. Schematic diagram of a flow of fluid under combined shear conditions: a — between flatly parallel plates under the action of pressure difference $\Delta P = P_1 - P_2$ (the upper plane moves in the direction transverse to the main flow); b — between two coaxial cylinders rotating towards one another at angular velocities Ω_1 and Ω_2

has been used and adequately developed in a number of works (see, for example, [27]).
Therefore, we may discuss here only some notes and quantitative assessments obtained recently, the importance of which is not purely theoretical but also essential for technological applications.

2.2.1 Estimation of the Fluid Elasticity's Contribution to the Flow-Rate (Velocity) Characteristics of Flow

Helical flow being analyzed as resultant from two independent flows (axial and circular), we may well assume that stable flow parameters (at least the flow rate) are determined primarily by viscous (flow) properties of the system, and the highelasticity effects (at superimposition of two flows) can be neglected in this case with a sufficient degree of accuracy which is reasonable from the point of view of engineering. The above assumtion was checked for correctness in [28,29] in a specific model of a viscoelastic fluid.

Flow in a circular channel with a significant relative length l/H (here l — is the length of circular head; $H = R_2 - R_1$ is the width of clearance, i.e., the difference between the inner radius of the outer cylinder, the tip, and the outer radius of the inner cylinder, the core) was simulated by the flow of a polymer between two parallel plates removed from one another to a distance H (see Fig. 1a). The resultant flow occurs due to the pressure difference $\Delta P = P_1 - P_2$ and motion of the upper plate with velocity U_0 in the direction transverse to the axial flow. In this case boundary conditions in the Cartesian system of coordinates are:

$$
\begin{aligned}
&\text{at} \quad y = 0; \quad v = 0; \quad U = 0 \\
&\text{at} \quad y = H; \quad v = 0; \quad U = U_0 \, .
\end{aligned}
\tag{1}
$$

The behavior of a non-Newtonian viscoelastic fluid can be described by a constitutive equation which takes into account condition (1). Rheological behavior of the fluid is described by an equation derived from White-Metzner-Litvinov model and takes the following form [27,32]:

$$
\sigma = -pi + \eta B_1 - 0.5\beta_1\beta_2 + \beta_2(B_1^2 + 0.5B_2^2) \, ,
\tag{2}
$$

where p is isotropic pressure; i is unit tensor; η is effective (apparent) viscosity; B_1 and B_2 are kinematic matrixes; β_1 is the coefficient which determines the value of normal stresses and equals, according to G. V. Vinogradov and A. Ya. Malkin (see, as example, [33]):

$$
\beta_1 = \eta^2/G_0
\tag{3}
$$

G_0 initial module of high elasticity; β_2 is coefficient of normal stresses taking into account secondary flows.

It is known [32,33] that $\beta_2 \ll \beta_1$ and, therefore value β_2 can be neglected, i.e., we take $\beta_2 = 0$. With respective values β_1 and β_2, constitutive equation (2) takes the following form:

$$
\sigma = -pi + \eta B_1 - 0.5 \, \beta_1 B_2 \, .
\tag{4}
$$

Using (4) to write stress tensor P_{xx} components as P_{yy} and P_{zz}, matrix components for the case under consideration are expressed specifically and are substituted into expressions for stress tensor components. Thus, we obrais the following relationships (details of transformation in [29]):

$$
\begin{aligned}
P_{xx} &= -0.5\beta_1(\partial v/\partial y)^2; & \tau_{xx} &= 0 \\
P_{zz} &= -p - 0.5\beta_1(\partial v/\partial y)^2; & \tau_{xy} &= \eta(\partial v/\partial y) \\
P_{yy} &= -p; & \tau_{xy} &= \eta(\partial u/\partial y)
\end{aligned}
\tag{5}
$$

The flow equation can be written in terms of stress components (in Cartesian coordinates) in the following form:

$$
0 = -\frac{\partial p}{\partial x} + \frac{\partial \tau_{xy}}{\partial y}
\tag{6}
$$

$$
0 = -\frac{\partial \tau_{zy}}{\partial y}
$$

With respect to (5), system (6) is rewritten as follows:

$$
\frac{\partial}{\partial y}\left(\eta\frac{\partial v}{\partial y}\right) = \frac{\partial P}{\partial x} = -\frac{\Delta P}{1} = -P = \text{const}
\tag{7}
$$

$$
\frac{\partial}{\partial y}\left(\eta\frac{\partial u}{\partial y}\right) = 0; \qquad \frac{\partial P}{\partial y} = 0
$$

Thus, the problem of flow of a viscoelastic fluid between two flat parallel plates one of which is moving in a direction transverse to the main flow is reduced to a solution of simplified system (7) at boundary conditions (1). Analysis of relationships (7) for specific boundary conditions indicates that the problem is reduced to the case of a non-Newtonian viscous fluid. In other words, the velocity profile v(y) is determined only by viscous characteristics of the media and the effect of high-elasticity properties of the melt upon velocity (flow rate) characteristics of the flow can be neglected.

2.2.2 Quantitative Analysis of Spiral Flow (Combined Shear) and Asymptotic Solutions for a Power-Function Fluid

Rheodynamics of non-linear viscous fluids flowing in circular channels with moving walls is described most comprehensively in [15, 34]. With respect to the above conclusion (see sect 2.2.1) that the high elasticity of a melt influences insignificantly flow rate parameters of a flow, the combined shear is discussed in [24, 28 – 30, 34] on the basis of a general approach to the analysis of viscosimetric flows developed by B. Colleman and W. Noll.

In the case usually considered (see Fig. 1) for a flow the following condition is observed:

$$
\begin{aligned}
P_{xy} &= \tau(\dot\gamma); & P_{xz} &= P_{yz} = 0 \\
P_{xx} - P_{zz} &= \sigma_1(\dot\gamma); & P_{yy} - P_{zz} &= \sigma_z(\dot\gamma)
\end{aligned}
\tag{8}
$$

where $\dot{\gamma}$ is shear velocity; p_θ — are stress tensor components; σ_1 and σ_2 are the first and the second difference of normal stresses.

It is convenient to consider the stable flow in a clearance between two infinitely long cylinders with radii R_1 and R_2, one of which is rotating with angular velocity Ω in a cylindrical system of coordinates. Each particle of material describes a curve along the common axis of cylinders z with angular velocity $\omega(r)$ and longitudinal velocity $U(r)$.

Velocity vector has the following components at each point: $v_r = 0$; $v_z = U(r)$; $v_y = \omega(r) \cdot r$ (here ω is angular velocity of the point, and r is the current radius).

Non-zero components of strain rate are written in the following way:

$$\varepsilon_{zz} = \partial u/\partial z = U'; \qquad \varepsilon_{\theta r} = (\partial \omega/\partial r) \cdot r = \omega' \cdot R \qquad (9)$$

At each point the local coordinate can be turned so that only one component of the strain rate vector is different from 0: $\dot{\gamma} = [(U')^2 + (\omega')^2 r^2]^{1/2}$. In the general case the vector of velocity at each point will not coincide with the vector of the rotated system of coordinates. Equation (8) is takes for rotated base at each point and rotated in the reverse direction to bring it in register with the initial position. Thus, we get the following relationships in the cylindrical system of coordinates:

$$p_{rz} = v\tau(\dot{\gamma}); \qquad p_{\theta r} = \mu r(\dot{\gamma}); \qquad p_{\theta z} = \mu \cdot v \cdot \sigma_1(\dot{\gamma})$$
$$p_{rz} - p_{\theta\theta} = \sigma_2(\dot{\gamma}) - \mu^2 \sigma_1(\dot{\gamma})$$
$$p_{zz} - p_{\theta\theta} = (v^2 - \mu^2)\sigma_1(\dot{\gamma})$$

where $\qquad v = U'/\dot{\gamma}; \qquad \mu = \omega^2 \cdot r/\dot{\gamma}$

The above relationships (10) have been derived by B. Colleman and W. Noll with the use of quite a different method [27]. They used system (10) to obtain the following relationships describing the flow between two infinite cylinders:

$$p_{rz} = \frac{b}{r} - \frac{r \cdot f}{2} \qquad (11)$$

$$p_{r\theta} = M/2\pi r^2 \qquad (12)$$

$$\int_{R_1}^{R_2} \left(\frac{b}{r} - \frac{r \cdot f}{2} \right) \frac{\varphi[S(\mu)]}{S(r)} dr = 0 \qquad (13)$$

$$S(r) = \left[\left(\frac{M}{2\pi r^2} \right)^2 + \left(\frac{b}{r} - \frac{r \cdot f}{2} \right)^2 \right]^{1/2} \qquad (14)$$

$$\Omega = \frac{M}{2\pi} \int_{R_1}^{R_2} \frac{\varphi[S(\tau)]}{r^3 S(\tau)} dr \qquad (15)$$

$$Q = \pi \int_{R_1}^{R_2} \left(\frac{r^3 f}{2} - rb \right) \frac{\varphi[S(r)]}{S(r)} dr \qquad (16)$$

Cylinder length l being sufficiently significant in relation to its diameter, value M is the moment related to cylinder length l; $f = p/l$ (here P is pressure at inlet to the clearance between the cylinders). At simple shear the flow function is $\varphi[S(r)] = \varphi(r)$; at $\tau < 0$, $\varphi(\tau) = -\varphi(\tau)$.

At combined shear the velocity of flow can be determined in a unique fashion with the help of the function $\varphi(\tau)$. Giving specific values of the moment and inlet pressure (M & f) with the help of equations (13)–(15), we can determine Ω and b and volume flow Q respectively, from (16).

In [24, 29, 34] some important properties of eqs. (11)–(16) were pointed out. It is assumed, without limiting the generality, that $R_2 = 1$ and $R_1 < 1$ while R_1 and R_2 are considered as dimensionless values of relative radii (normalizing, for example, on the basis of R_2). Taking into account that $\partial\varphi(S)/\partial S \geq 0$, we may assume that the integral

$$ J = \int_{R_1}^{R_2} \left(\frac{b}{r} - \frac{rf}{2} \right) \frac{\varphi[S(\tau)]}{S(\tau)}\, dr $$

is a strictly increasing function and $b(dJ/db) \geq 0$. Since J, for b equal to $(R_1 \cdot f)/2$ and $f/2$ respectively, changes sign, eq. (16) can have only one solution for each R_1, M and f. Proceeding from the condition that the flow strictly increases, $\varphi(S)$, it follows that $(\partial J/\partial f) < 0$. Thus:

$$ \frac{\partial b}{\partial f} = \frac{\partial J/\partial f}{\partial J/\partial b} > 0 $$

and b is a strictly increasing function of f. Since the second derivative of the function is $[(b/r) - rf/2] > 0$, and $[\varphi(S)]/S$ is also a strictly increasing function, we get:

$$ b < \left(\frac{R_1 + 1}{2} \right)^2 \cdot \frac{f}{2} $$

Thus, the maximum longitudinal velocity is:

$$ U(r) = \int_{R_1}^{r} \left(\frac{b}{r} - \frac{r \cdot f}{2} \right) \frac{\varphi[S(r)]}{S(r)}\, dr $$

attained at $r < (R_1 + R_2)/2$. It follows from the bounded character of function

$$ \left| \frac{b}{r} - \frac{r \cdot f}{2} \right| < \frac{f}{2} \left(\frac{1 - R_1^2}{R_1^2} \right) $$

within possible range of changes in b(f, M) and from condition $[\varphi(S)/S]' > 0$ of Eq. (15), that $\Omega \to \infty$ for rotating cylinder at $M \to \infty$. If $[\varphi(S)/S]'' > 0$, it follows from the expression of the flow velocity (flow rate) written as $Q = 2\pi \int_{R_1}^{1} U(r)\, r^2\, dr$, that with increasing M flow velocity also increase (Q = const for a curve of a flow with the second Newtonian viscosity at M → ∞).

For the case of approximation of $\varphi(S)$ by power function (the known Ostwald-de Villes model) $K\tau^n$, when $n \geqslant 0$ and $K > 0$, it follows from Eq. (16) that for $b \to b_0$, $M \to \infty$:

$$b_0 = fR_1^2 \frac{n-1}{2(n-2)} \cdot \frac{1 - R_1^{2(n-2)}}{1 - R_1^{2(n-1)}} = fK(n, R_1) \tag{17}$$

and beside that $(b - b_0) \sim 1/M^2$ and does not change sign at fixed n.

It can be seen from (15) and (16) at M → ∞ that

$$(\Omega/\Omega) \to 1 \quad \text{and} \quad (Q/Q_0) \to 1,$$

where

$$\Omega_0 = \frac{K}{2n} \left(\frac{M}{2\pi}\right)^n \cdot \frac{1 - R_1^{2n}}{R_1^{2n}}, \tag{18}$$

$$Q_0 = \pi K \left(\frac{M}{2\pi}\right)^{n-1} \left\{\frac{f}{4(n-3)}\left[\frac{1}{R_1^{2(n-3)}} - 1\right] - \frac{b_0}{2(n-2)}\left[\frac{1}{R_1^{2(n-2)}} - 1\right]\right\}. \tag{19}$$

Numerical solution of equations (13)–(19) for polypropylene extrusion was made in [29, 34] using approximation of the flow function (flow curve) by a piecewise power function. To find the root of b(f, M) of Eq. (13), the authors used a formal search algorithm compiled as a standard program for computer M-20 (USSR). Figure 2 gives dependency of b/f upon M (M is the specific moment of a core's rotation, i.e., the moment related to the length of the channel). It can be seen in Fig. 2 that (b/f) is a strictly decreasing function.

The next figures give curves of dependency of M upon angular velocity of rotation Ω (Fig. 3), the strictly increasing functions; dependencies of bulk velocity of the flow $Q(\Omega)$ (Fig. 4) for different values of $f = p/l$ (*continuous lines* correspond to calculations of Eqs. (13)–(16) and *dotted lines* correspond to similar calculations with

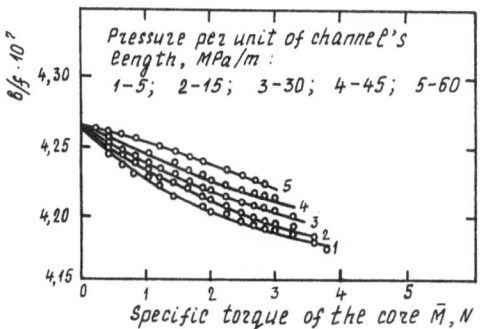

Fig. 2. Parameter (b/f) versus specific torque M of the core at different pressure per unit of channel's length, MPa/m: 1–5; 2–15; 3–30; 4–45; 5–60

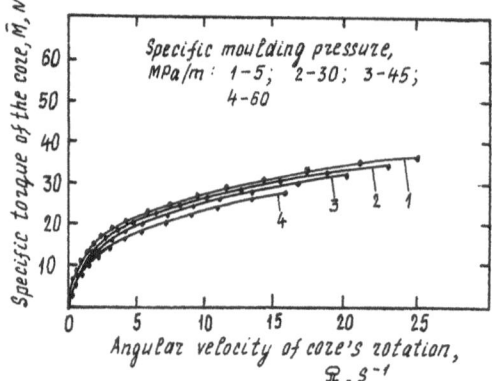

Fig. 3. Core torque related to channel length versus angular velocity of core rotation (Ω) at different specific molding pressures (MPa/m): 1–5; 2–30; 3–45; 4–60

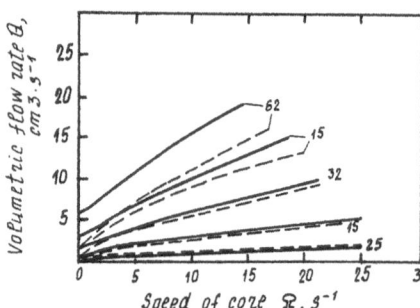

Fig. 4. Volumetric flow rate of molten polypropylene versus angular velocity of core rotation at different values of specific pressure difference f, (MPa/m) specified on the curves. *Continuous curves* have been calculated from equations (11)–(16) using numerical methods, *dashed curves* have been calculated from asymptotic formulae (17)–(19)

asymptotic assumptions for Eqs. (17)–(19)[1]. Value f, for which approximation expressions (17)–(19) can be used, has been estimated proceeding from inequality

$$f < \frac{M/2n}{K - {}^{1}/_{2}} .$$

(20)

It should be noted that even in the cases when $f = (M/2n)/(K - 1/2)$ the approximated formula gives a satisfactory result[2]. The error in approximation of flow rate value Q, compared to the results obtained with the help of numcerical methods, turns out to be less than 20%. It should be noted that the authors have described a method used successfully for calculation of Q the core being at rest when M = 0, see, for example [35].

Distribution of effective viscosity values of molten polypropylene over a cross-section of the circular clearance is given in Fig. 5. Viscosity varies within the following limits: the upper curve corresponds to a flow with motionless core (ω = 0, M = 0)

---

[1] Approximated values of Q were calculated [24, 29] for a specified M and r (which were equivalent to R_1 and R_2), then the values of function [S(r)] were determined proceeding from (14). The second member in the right-hand part of (14) is rather insignificant, so it was neglected. Using the calculated value S, function φ(S) was approximated by a straight line in logarithmic coordinates which allowed the determination of the values n and K (parameters of power function) used further in Eqs. (17)–(19).
[2] For example, at M = 1.5 and f = 4.5, relationship $[f(K - {}^{1}/_{2})/(M/2n)] > 1$ [28, 34] is true.

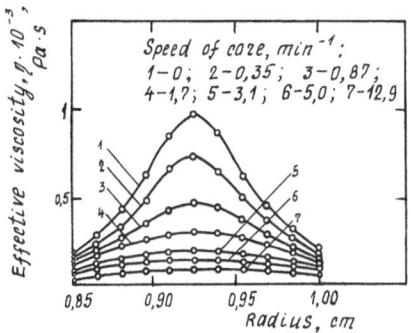

Fig. 5. Distribution of effective viscosity values in molten polypropylene over cross-section of the circular molding slit at specific pressure in the head 45 MPa/m and different core speed (min^{-1}): *1* — 0; *2* — 0.35; *3* — 0.87; *4* — 1.7; *5* — 3.1; *6* — 5.0; *7* — 12.9

and the lower one corresponds to the melt flow in a narrow clearance between two concentric rotating cylinders.

Total specific power consumed (related to cylinder length l) determined theoretically as $N = f \cdot Q + \omega M$, is given in Fig. 6 as a function of the core's speed of rotation Ω. Note that power required for extrusion N_{ext} and power required for rotation of the core N_{rot} are strictly increasing functions of ω at a fixed specific pressure difference $f = p/l$. Power consumed for extrusion N_{ext} at constant flow rate Q decreases with increasing pressure difference while N_{rot}, under the same conditions, remains a strictly increasing function of Ω. In conclusion, it should be noted that these theoretical assessments of power consumption suffer certain changes when we come to practical evaluation of the real process of extrusion molding (the reasons are given in Sect. 1.3); here we may only point out that in practical experience emphasis is placed on the specific power consumption, i.e., power related to a machine's output.

Experiments carried out on laboratory and pilot plants, both with viscoplastic systems [13-15] and with thermoplastics, polyolefins (polyethylene, polypropylene, and filled materials based thereon) [7,9,28] and various PVC-compositions [36] corro-

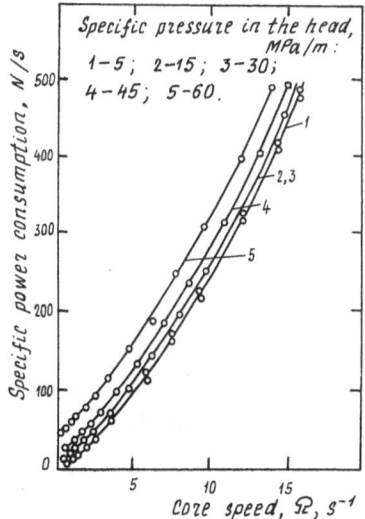

Fig. 6. Total theoretical power consumption for extrusion of molten polypropylene and core rotation in the head as a function of core's speed at different specific pressures in the molding head: *1* — 5; *2* — 15; *3* — 30; *4* — 45; *5* — 60

borate the satisfactory accuracy of extrusion flow rate characteristics according to the theory discussed above. Figure 7 exemplifies data [28, 29] on the dependency of flow ($cm^{-3} \cdot s^{-1}$) through an angular circular head (Fig. 8) during extrusion of polypropylene upon angular velocity of core rotation (rad/s) at different values of pressure difference produced by the screw (laboratory plant based on single-screw extruder ET-20 with a screw diameter D = 20 mm and relative length L/D = 20:1 manufactured by Anger-Plastik-Machinen, Austria). The experiment (*the points*) coincides with a high degree of accuracy with the theory (*continuous lines*).

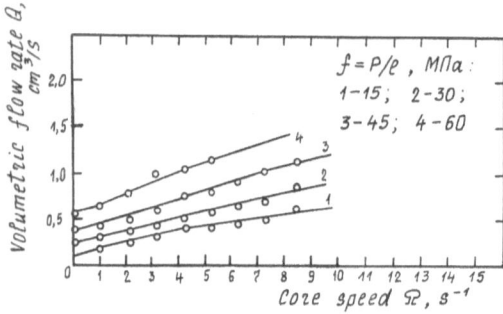

Fig. 7. Experimental dependencies of volumetric flow velocity of molten polypropylene upon speed of core's rotation in a circular head for different values of f (MPa/m): *1* — 15; *2* — 34; *3* — 45; *4* — 60

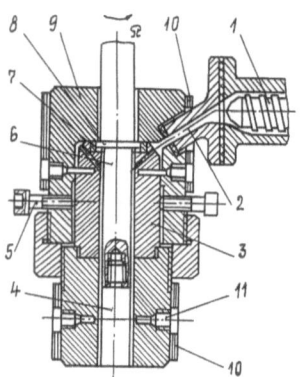

Fig. 8. Design of angular circular extrusion head with rotating core for manufacturing of tubes and hoses: *1* — Single-screw extruder; *2* — Adapter; *3* — Core holder; *4* — Separable core tip; *5* — Alignment bolts; *6* — Rotating core; *7* — Plain bearing; *8* — Stop flanging; *9* — Head casing; *10* — Electric heaters; *11* — Place for installation of thermocouples

2.3 Other Principal Patterns Employed to Produce Combined Shear in Molten Polymers

a) There are several design approaches to the realization of the core's rotation. One of them, given in Fig. 8, employs an angular circular head similar, in principle, to the known molding tools used for application of cable insulation, extrusion of reinforced hoses or blanks for subsequent manufacturing of solid items (bottles and other types of hollow containers) (see [36–38] by extrusion-blow molding method).

In all cases this variant (Fig. 9a) implies an independent core actuator rotating in a plane perpendicular to the main flow (direction of extrusion).

b) Another approach (Fig. 9c) realized, for example, in [36], employs a core in-

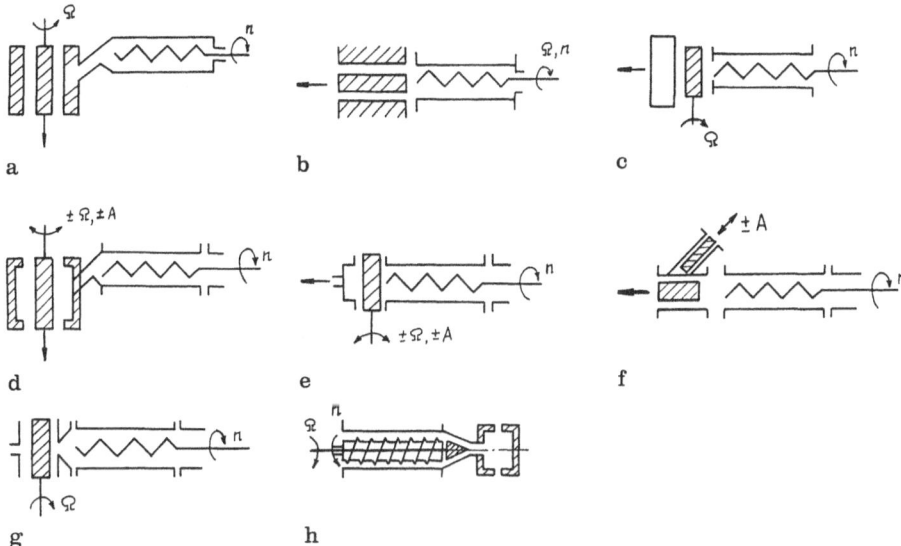

Fig. 9a–h. Different design patterns for superimposition of rotation, reciprocating-rotary and oscillating vibrations on the main flow of molten polymer under extrusion (a–d) and pressurized moulding (e–h). Explanation of flow diagrams is given in the text.

n = speed of screw rotation; Ω = angular velocity of core rotation or rotation of an element in the zone upstream of the molding channel; $\pm\Omega$ and $\pm A$ is angular frequency and amplitude of reciprocating-rotary (oscillating) vibrations of the core or the element in the zone upstream of the moulding channels

stalled in a once-through circular head as an extension of the extruder screw (it screwed in into the core instead of the tip). In this case the core does not have an independent actuator and its speed of rotation coincides, naturally, with the speed of rotation of the screw itself. The advantage of this design is that it is simple and excludes the second drive unit, but there are also significant drawbacks: the absence of technological "flexibility", impossibility of adjusting the speed of core rotation independently of the screw. Similar designs can be used for extrusion of tubes and hoses. In this case the core will rotate perpendicularly to the flow generated by the screw and a helical (spiral) flow occurs in the circular clearance of the head.

c) Several types of head design have been developed [29,39] in which the rotating element is mounted transverse to the central line of extrusion and the direction of its rotation coincides with that of the flow (Fig. 9c). Here the rotating element has an independent actuator but the flow in the head is not helical; nevertheless, the type of strain in the melt can be classified as combined shear since there is 'superimposing' of an additional shear upon the main flow (i.e., additional shear caused by the rotating element). The additional shear direction coincides here with that of the main shear. Experiments carried out with such heads [29,39] have confirmed the merits of molten thermoplastic molding under combined stress conditions but have not revealed any advantages compared to designs described in a and b.

d) Beside the rotating elements, cores mounted according to a and b can perform, with thel help of drives fitted with eccentric mechanism, reciprocating-rotary oscillations (vibrations) at a low frequency and high amplitude (Fig. 9d). Thus, in [29, 30] the author described experiments in the extrusion of polypropylene at reciprocating-rotary vibration of the core mounted according to design a with a cyclic amplitude (by the angle of turning) of 4.8°, 11.5° and 22.3° at a frequency of 25 Hz. In this case the core vibration caused cyclic shear strains of 75, 180, and 335%, respectively. Vibrations with closely spaced parameters in longitudinally and laterally mounted elements in the heads designed according to a and c were generated during the extrusion of polyethylene [39] and polypropylene-based compositions with different aggregates (chalk, soot, asbestos) [9, 28, 31].

e) Vibration effect upon a melt can be superimposed in a plane close to the main flow with the help of the so-called oscillating cores (Fig. 9e). According to the theory, the combined shear occurs also in this case as a combination of two shear strains having practically one and the same direction coinciding with the central line of extrusion. The theory of such flows has been also discussed by Colleman and Noll in the above-mentioned publications [17, 41]). L. A. Feitelson's fundamental work [42] focused on the oscillating effect upon molten polymers and a thorough investigation into the effect of the so-called vibrothyxotropy in pure and filled polymers.

f) A number of designs employing the rotating (longitudinally and laterally to the main flow) and oscillating elements in the zone of once-through extrusion heads installed upstream of the inlet to molding channels shaped differently has been described and realized practically in laboratory and pilot plants [23, 29, 39, 40]. These designs (Fig. 9d) have no limitations in terms of shape (cross section) of the manufactured semifinished products and can be recomended even for granulating heads [23].

g) The rotating element, the torpedo, for injection assembly of a casting machine (Fig. 9h) was developed, apparently, for the first time by the authors of invention [43]. This design achieved in the laboratory thermoplastic machine an increased injection rate of 20 to 30%, reduced temperature by 15 to 20 °C and injection pressure by 10 to 15%, whileproviding a possibility of performing directed adjustment of rheological properties up processed thermoplastics. Simultaneously, minimum readjustment of injection conditions is reduced when a batch of material is changed for another one (even when the type of polymer is changed, for example, polystyrene for polypropylene) without any change in the molding nozzle design; it has not found, unfortunately, a commercial application. Probably this is related to the rapid progress in other spheres of improvement in casting machines vital for their capacity and the quality of manufactured products. These spheres include intensification of the casting cooling process, increasing the accuracy of setting and maintaining temperature in the zones of the plasticizing assembly and the mold, automatic and computer-aided control of all systems (hydraulic, mechanical, thermal) and operation which became in the 1970s and 1980s major lines of progress in the sphere of pressurized casting.

Other areas of application of vibration technology in casting molding processes include, for example, vibroplasticizers- the designs of which have been developed by Western companies (the so-called teledynamic machines manufactured by Engel (Austria), Bücher-Güyer (Switzerland), Braas (FRG), and other) and Soviet organizations and are described in detail in [10] and leaflets of manufacturers.

2.4 Major Regularities of the Effect of Rotation and Core Vibration upon Rheological Properties and Process Parameters of Polymer Extrusion

In Sect. 2.3 we gave a brief classification of the major designs employed to generate combined stress conditions of melts in molding tools, the heads and in channels upstream of them, through rotation and vibration of cores. It should be noted that strict and sufficiently comprehensive works providing a comparative quantitative evaluation of all the abovementioned methods and designs have not been published yet. Nevertheless, we may consider that general qualitative tendencies of the effect of rotary and vibration motion of the molding elements of heads upon rheological characteristics and process parameters of thermoplastic extrusion have been identified and substantiated. The common character of these tendencies, independent of the pattern of vibration superimposition and location of the moving element in the extrusion head, has been corroborated by analysis. Certainly, the extent to which different effects, which we shall discuss below, are displayed depends significantly on the design, the zone of vibration effects (at inlet to the head or in molding channels), nature and molecular-mass characteristics of the polymer, presence of a filler, rate of extrusion, etc. With irreversible changes in the material caused by thermomechanical destruction being excluded, the mechanical (low-frequency and vibration) effects relax while the rate and completeness of the relaxation processes are determined exactly by the abovelisted factors.

In certain works [28, 29, 36, 39] authors discussed the effects occurring in rotation and vibration of elements installed in extrusion heads in a different manner (along or transverse to the main flow, at inlet to the molding part of the heads or in molding channels) but no principal differences causing one to "insist" on a specific pattern or design to be taken alone have been revealed in case of comparatively short molding channels [44]. In case of comparatively long channels (see 1.4(a)) many effects can "relax" partially or even completely and vibration effects can be levelled ("restored" thyxotropically).

2.4.1 Effective Viscosity, Flow-rate and Pressure Characteristics of the Extrusion Process

Effective Viscosity, Flow-rate and Pressure Characteristics of the Extrusion Process change notably during the above-discussed movements of the core in the extrusion head. Figure 10 gives tentative "flow curves" (flow-rate/pressure characteristics) determined according to the procedure described in [28, 9] for processing of polypropylene with melt flow index MFI = 0.5 g/10 min (at 230 °C under a load of 21.6 N appr.) the core is rotating in an angular head with a circular channel (according to design shown in Fig. 9). Data indicate that curves shift, upwards and to the left as the speed of core rotation increases. This effect is more pronounced in polypropylene than in polyethylene (compare, for example, to data of [22]). The observed decrease in effective viscosity of a melt compared to the usual extrusion with a core at rest (when $\Omega = 0$) results in an increased flow rate at a preset pressure difference (ΔP) in the head when the rotation of the core is superimposed at constant extrusion temperature, or, at a constant flow rate, the pressure required for extrusion of the material drops significantly. Reduction in the apparent (effective) viscosity can be explained

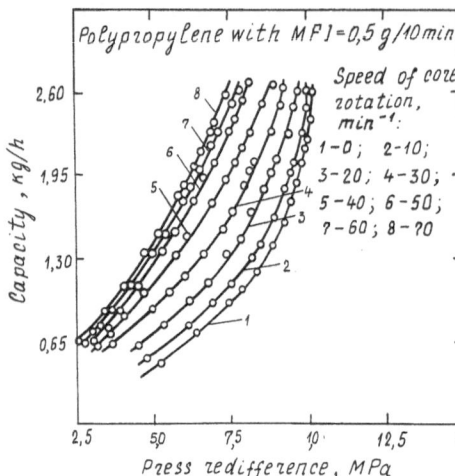

Fig. 10. Pressure/flow-rate characteristics ("tentative flow curves") of angular circular extrusion head (the design is given in Fig. 8) in molding of polypropylene with MFI = 0.5 g/10 min under conditions of core rotation at a speed of, min^{-1}: $1 — 0$; $2 — 10$; $3 — 20$; $4 — 30$; $5 — 40$; $6 — 50$; $7 — 60$; $8 — 70$

in the following way. It is well known that viscosity of anomally-viscous (non-New-tonian) liquids is determined by the second invariant of the kinematic stress tensor (J_2), values of η_{eff} drop as the value of J_2 increases [15, 27, 32, 33]. Increase in J_2 can be attained either due to increased gradient of shear velocity $\dot{\gamma}$ of the axial flow, or by superimposing of the second rotary motion on this flow. Reduction in η_{eff} at a spiral flow results in increased velocity (flow rate Q) of the main flow at ΔP = const. In cases with Q = const superimposing of the circular motion of the core allows a reduction of the required value of ΔP, as well at the counterpressure on the screw 'unloading' thus the thrust bearing of its drive. Two facts should be noted here: firstly, mutual confluence of the axial and circular flows increases in media with a more pronounced anomality of viscosity (a similar effect was observed many times in viscoplastic polymeric and non-polymeric materials [6, 13–15]); secondly, reduction in effective viscosity of the media slows down and then stops completely at a suffi-ciently high speed of core's rotation (in the case illustrated by diagrams in Fig. 10, beginning with Ω = 50–70 min^{-1}). This can be explained, apparently, by the fact that as the average linear velocity of melt's flow in the channel increases, a moment comes when the time which the medium stays in the circular channel becomes comme-asurable (and then even less) with the time required for the particles of the apparent viscosity becomes again higher than that calculated from the spiral flow theory. This phenomenon is corroborated indirectly by the convergence of curves in Fig. 10 at Ω > 50 min^{-1} (curves 6, 7, 8) and has been observed earlier by investigators [6, 7, 14, 15].

Value of η_{eff}, Q, and ΔP similar in terms of tendencies and scale of changes have been confirmed experimentally for polyethylene and polypropylene at low-frequency oscillations of the core (25 Hz) at amplitudes up to 22.3° [29, 39].

Rotation of the core (or its reciprocating rotary vibration) can be even more efficient in processing of high-viscous melts, for example, filled polymers, high- and superhigh-molecular polyethylene (with MM $\geq$ 10^6). We may assume that this is dependent upon two major causes. The introduction of a filler results in a changed spectrum of relaxation time H(θ) [41, 42, 45]. Thus, for example, introduction of 10% of chalk (by volume) into polyolefins "shifts" the spectrum along the axis of coordinates towards higher values of H while the form of the spectrum remans unchanged. However, as filler quantity increases up to 15% (by volume), the spectrum elongates a big with the 'shift' in the direction of longer relaxation time (θ). This alteration of the relaxation spectrum results, in its turn, in a significantly higher viscosity (since the maximum Newtonian viscosity (η_0), for example, is assessed as an integral characteristic

of the spectrum: $\eta_0 = \int_0^\infty H(\theta)\, d\theta$. Formation of the combined stress in a melt reduces

'deformation' of spectrum and viscosity of the filled system. Another cause of reduced viscosity is the destruction of the 'framework' formed by the filler itself when its content by volume is sufficiently high with rotating and oscillating moulding elements. Certainly, both the above mentioned causes of the favourable effect of the spiral flow and cyclic strains upon the viscosity of filled polymers are interconnected. Easier processing of filled thermoplastics due to core rotation was observed, apparently, for the first time in [9, 31] in polyethylene and polypropylene filled with 10% (by mass) of chalk and 20% (by mass) of asbestos as illustrated by curves given in Fig. 11.

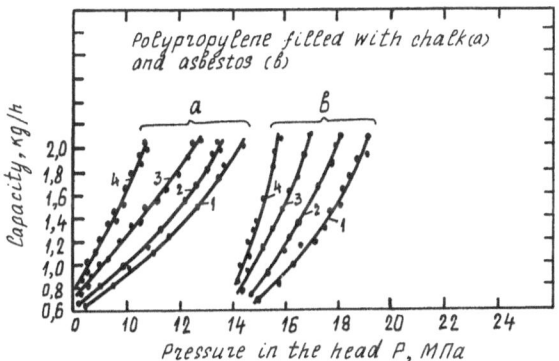

Fig. 11. Pressure/flow-rate characteristics of extruder (capacity versus pressure drop across the head) in processing of polypropylene filled by 10% (by mass) of chalk (*a*) and 20% (by mass) of asbestos (*b*) at a temperature in the head equal to 210 °C and amplitude of reciprocatingrotary vibration of the core, degr.: *1* — 0; *2* — 4; *3* — 11.5; *4* — 22.3

Rotation of the core at $\Omega = 70$ min^{-1} shifted flow rate/pressure characteristics of extrusion curves upwards and to the left, the result of which was a notable drop of $(\Delta P)_{70}$ compared to $(\Delta P)_0$ (indexes used here correspond to the speed of core's rotation). Increase in $(\Delta P)_{70}$ up to values of $(\Delta P)_0$ due to higher speed of the screw allowed as increase is the extruder's capacity by 40–80% (depending on the nature and content of the filler). Results very much similar to that were obtained later in experiments with filled PVC-based compositions [36]. Pressure in the head can be reduced by not less than 20–30% at a constant extrusion rate.
experiments with filled PVC-based compositions [36]. Pressure in the head can be reduced by not less than 20–30% at a constant extrusion rate.

Experimental data given in Fig. 12 are interesting both from the practical and theoretical point of view. The figure gives tentative "flow curves" of high-pressure (low-density) polyethylene with MFI = 2.0 g/10 min at a temperature of 170 °C in chan-

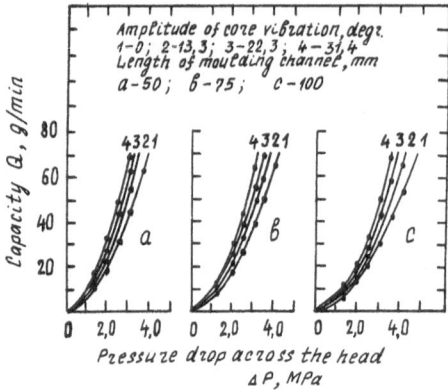

Fig. 12. Tentative "flow curves" of low-density polyethylene with MFI = 2.0 g/10 min extruded at 170 °C through channels with a two-angle ellipse Wber cross section with a length of 50 (*a*), 75 (*b*), and 100 (*c*) mm with reciprocating-rotary vibration of the element in the zone upstream of the inlet to the channel (according to the diagram given in Fig. 9);
1 — The core is motionless; *2*, *3*, and *4* — the core vibrates at a frequency of 25 Hz and amplitudes (degr.): 13.3 (*2*), 22.3 (*3*), and 31.3 (*4*)

nels of different length with a two-angle ellips cross-section (the so-called Weber cross-section) at oscillating motion of the element installed in the head according to the design shown in Fig. 9d — upstream of the inlet to channels. Several conclusions can be made:

a) A polymer 'memorizes' well vibration effects applied to it prior to admission to the capillary; these effects do not have time to relax despite the rather significant (50–100 mm) length of the channels.

b) Amplitude (A) of reciprocating-rotary vibration being increased from 0° to 31.3° increases spinneret's capacity by 70–80% but further elevation of A-values is iseffective, values η_{eff}, ΔP, and Q remain practically unchanged.

2.4.2 High-Elastic Properties of Molten Polymers and Filled Compositions

High-Elastic Properties of Molten Polymers and Filled Compositions also suffer notable changes. Several works have been published in which it is exemplified by characteristics very important for the extrusion technology, such as the values of critical flow parameters and swelling factors [9,39,44]. Rotation of the core in the molding part of the head results in increased critical velocities $\overline{D}_{cr}$ (i.e., velocities of shear at which the so-called elastic turbulence conditions [33,46] of extrusion (or unstable flow) occur (Fig. 13). Thus, the range of "smooth" extrusion velocities is extended within which the produced extrudate is free of form distortion, surface roughness and ruptures. At the same time, manifestations of the Barrus effect decrease: post-extrusion swelling factor of jet K decreases (for round channels $K = d_e/d_K$, and for channels with a complicated non-round cross section $K = (S_e/S_K)^{1/2}$ where indexes "e" and "K" belong to the extrudate and the channel, respectively, while d and S are diameters and cross-section areas). Thus, for example, if at $\Omega = 0$ for poluolefins K = 1.7–2.0, at $\Omega = 70$ min^{-1} K = 1.4–1.5. Reduction in value K

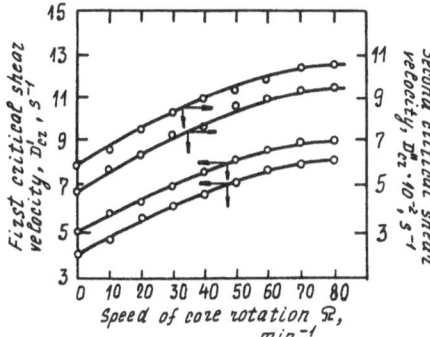

Fig. 13. Effect of core rotation according to the pattern given in Fig. 9a upon critical shear velocities in polypropylene: D'_{cr} — the first critical velocity at which minor roughness occurs on the surface of the extrudate; D''_{cr} is the second critical velocity at which the extrudate features large-scale distortions, corrugation, roughness

contributes to a higher stability and accuracy of dimensions of the produced extrudates. At constant temperature and pressure in the head, cyclic strains of the core reduce swelling of polypropylene approximately to a similar extent: from 1.9–2.0 at A = 0 to 1.5 at A = 22.3°.

The effect of the spiral flow and vibration on the abovementioned characteristics of melts ($\overline{D}_{cr}$ and K) are accounted for by the effect of the combined shear on the relaxation characteristics of the material [9,41,42] and on the orientation of macromolecules of the melt in the extrusion head [9,30,34]. A study of molecular orientation of polypropylene in extrudates (employing quantitative evaluations according to the infrared spectroscopy procedure described, for example, in [30,47]) has revealed that the superimposing of the core's rotation reduces orientation effects in the flow direction and, respectively, accumulation of high-elastic strains. The latter determine the occurrence of unstable flow and jet swelling. On the other hand, an opposite tendency has been noted in extrusion of PVC [36]: swelling of extrudate increases with speed of core's rotation (to a certain limit). This effect is characteristic only of PVC and has not been described for other thermoplastics. It, apparently, depends upon the specific 'nodular' structure of PVC, the melt of which is not a 'true' one (at the molecular level) while it preserves peculiar supramolecular kinetic units, the "nodules", and the melt's flow is not realized according to the molecular mechanism as is the case of usual thermoplastics. Nodules preserved in molten PVC even at relatively high temperatures (about 193 °C) and shear velocities are practically incapable of accumulating high-elastic strains, therefore, the swelling factor of PVC-extrudates is comparatively low. Temperature rise and mechanical effects (rotation or vibration) of the core promote, apparently, the destruction ('melting') of PVC nodules effectuating thus a normal (molecular) mechanism of its flow at which the orientation of macromolecules increases and causes a rise of the swelling factor K. As regards the elastic turbulence, or unstable flow, one of the main causes of which is the significant tension of the melt in the zone upstream of the moulding tool [33,35,46,48,49], the positive effect of vibration upon the melt in the above-mentioned zone is observed also in PVC which allows one to obtain a smooth glossy extrudate even at significant rates of extrusion [36].

2.4.3 Physical and Mechanical Characteristics

Physical and mechanical characteristics of extrudates obtained from different thermoplastics at low-frequency cyclic effects upon melts have not been studied thoroughly
so far, although in some publications which have treated this matter earlier it was
mentioned that the above characteristics "at least do not deteriorate and are improved
in some cases significantly". However, the results of the recent works permit one to
arrive at more accurate and optimistic conclusions.

To exemplify this, we may refer to studies in which different patterns of vibration
superimposing on thermoplastic have been employed. Typical data obtained by
M. L. Fridman et al. [29, 39] for polyethylene at reciprocating-rotary vibration of an
element installed in the extrusion head transverse to the main flow are given below.
Vibration was superimposed according to procedure given in Fig. 9d upstream of
the inlet to molding channels which had a cross-section in the form of a twoangle
ellipse (Weber cross-section) with an area of 39.7 mm². On the example of low-density
polyethylene of a typical film brand (MFI = 2 g/10 min at 190 °C and at load of
21.2 N) extruded at 170 °C it has been demonstrated that the strength of obtained
extrudate increases with amplitude of circular displacement of the vibration element:

Table 1.

Moulding channel length, mm	Ultimate strength (σ_p) of extrudate at diffe- rent amplitudes of vibrating element's oscilla- tion in the head, grad			
	0	13.3	22.3	31.3
50	14.2	17.8	20.6	21.8
75	18.2	20.6	21.6	24
100	17.0	18.4	20.6	22.2

Results in much the same way have been obtained also by M. L. Fridman [28–30]
for polypropylene. We may assume that the strengthening of extrudate is related to
the increased orientation of macromolecules of polyolefins in a melt which is promoted
by reduction in viscosity at combined shear and the possibility of reduced extrusion
temperature in this case. At a constant length of the channel, strength grows with
amplitude (A) of vibration to a certain value (in specific experiments it was up to
31.3°) after which the dependency σ_p(A) is "limited". Note that σ_p grows first, as the
length of the molding tool increases, (compare to the above-mentioned data on σ_p
for channels with l = 50 and 75 mm, respectively) and then drops, but in all cases σ_p
of samples obtained uder conditions of vibration is higher than in the case without
vibration. We may assume that in the beginning the elongation of channels contributes to the growth of shear stresses and longitudinal orientation of the melt which
is fixed, to some extent, in the extrudate. In excessively long channels the orientation
relaxes significantly. This has been proven by quantitative measurement of orientation
in polypropylene samples done using the infrared spectroscopy method (ISM) according to a procedure developed by Yu. V. Kissin et al. [47, 50]: orientation of poly-

propylene increased up to 47–51 % under conditions of extrusion with superimposed core vibration compared to 33–36% when the core was at rest ($\Omega = 0$, A = 0). This caused a notable growth of σ_p of the extrudate in the longitudinal direction. However, the effect was observed in cases when the molding channel was "not too long".

Strengthening of extruded bars and tubes made of PVC-compositions at low-speed rotation and low-frequency vibration of the core in the head proven by M. M. Genender [36] is caused primarily, we believe, not by orientation effects but by increased homogeneity of multicomponent systems.

Not long ago the matter of efficiency of longitudinal and lateral vibration in extrusion was studied and reported by G. Casulli, G. Klermont, A. von Zigler, and B. Mena [51] in their report "Extrusion of polymers through oscillating heads". The authors studied the results of the use of longitudinal and lateral vibration in extrusion of molten polymers within a wide range of frequencies and amplitudes and corroborated earlier conclusions that superimposition of oscillations improves significantly the mechanical characteristics of extruded polymers. In some materials the authors [51] found an increase in strength by 150%, significantly increased torsional rigidity, reduced swellint of the jet and increased velocity of the flow. Such a ligh tensile strength and torsional strength are surely important for technology as well as other effects, although these were discovered experimentally and published, as mentioned above, much earlier by other authors, mainly by Soviet investigators.

2.4.4 Estimation of Power Consumption and Power Efficiency

Estimation of power consumption and power efficiency of the use of moving moulding elements is important both from the theoretical and the practical point of view. It is also rather complicated. In the end of Sect. 2.2 we stated that the theoretical power consumed for extrusion of a melt through the head and that consumed by rotation (or vibration) of the core are strictly increasing functions of frequency (ω or Ω) at a fixed specific pressure drop across the f = P/l (see Fig. 6).

However, this logical conclusion does not reflect the real 'technological' picture if we consider specific power consumption, i.e., consumption of power per unit of production (for example, per each kg of extruded mass) since it is precisely the index $q = N/Q$ (here N is power consumed, Q is capacity in kg/h) that characterizes the efficiency of the process and equipment. Fig. 14 gives power consumed for rotation of the core (a), total (b) and specific (c) power consumption polypropylene tube and hose extrusion plant versus output measured in [28–30]. The reason is that oower consumption it case of a melt flow in extrusion head $N \approx fQ + \Omega M$ account only for a part of total power consumed by the extrusion plant since a significant part of energy is spent, in particular, on the rotation of the screw. Rotation of the core, as mentioned above, reduces significantly the effective viscosity of the melt and pressure in the head at the same output of the process. This is accompanied by a decrease in the counterpressure on the screw and, consequently, reduced friction losses in mechanical elements (thrust bearings) of the extruder. The result is that in practical experience the rotation of the core can reduce power required to rotate the screw so that the growth of total power consumption slows down. Starting from a certain volumetric (Q) or mass (G) output, curves N(G) obtained at core rotation take

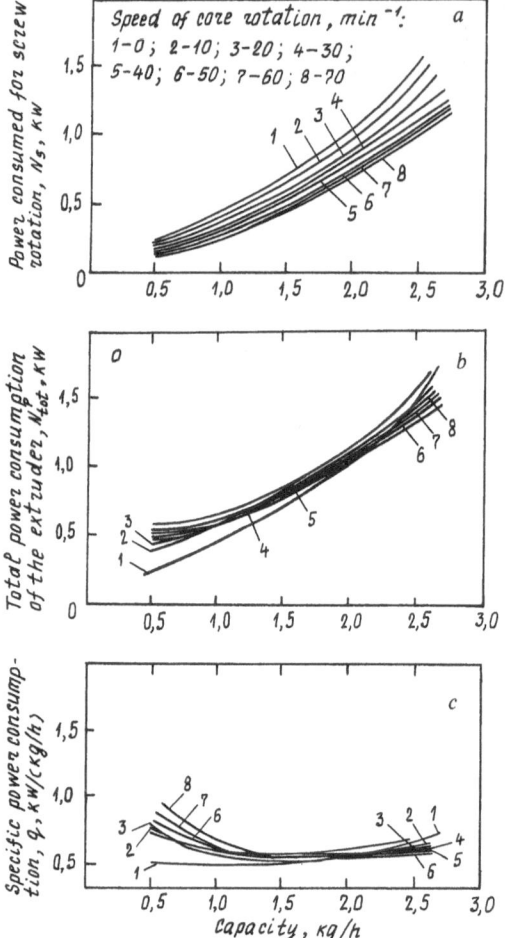

Fig. 14a–c. Effect of core rotation on the power consumption of the extrusion plant manufacturing polypropylene tube blanks: a — power consumed for screw rotation (N_s); b — total power consumption of the plant (N_{tot}^p); c — total power consumption calculated per unit of product's mass (q). Speed of core rotation, min^{-1}; $1 - 0$; $2 - 10$; $3 - 20$; $4 - 30$; $5 - 40$; $6 - 50$; $7 - 60$; $8 - 70$

positions, due to their minor slope (low rate of N's growth), ever lower than the curve of power consumption at conventional extrusion when the core is motionless [29, 30]. This tendency is similar to, but expressed more clearly, in specific power consumtion, or power per unit of production. At significantly high values of G extrusion with rotating molding elements will not increase power consumption but become advantageous from the point of view of energy.

A similar conclusion can be made from the results of the estimation of specific power consumption by an extruder outfitted with in a vibrating core (Fig. 15) employed for processing of polypropylene filled with chalk and asbestos [31, 42, 44]. Specific power consumption is higher in case of the vibrating core, however, the output of the process is also increased and curves given in Fig. 15 converge gradually so that at high rates of extrusion the capacity of the machine can be increased practically without any growth of the specific power consumption. This is more characteristic of filled compositions, in particular, as given in Fig. 15b, in case of polypropylene with 20 % (by mass) of asbestos when specific power consumption, beginning with

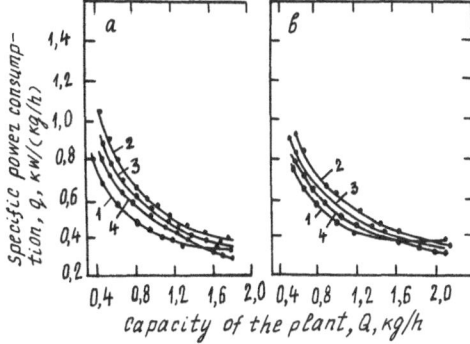

Fig. 15. Specific power consumed in the process of extrusion of polypropylene filled with 10 % (by mass) of chalk (a) and 20 % (by mass) of asbestos (b) versus capacity of the plant at core vibration at amplitudes, degr.: $1 - 0$; $2 - 4.8$; $3 - 11.5$; $4 - 22.3$

certain values of Q (or G), becomes even lower than in the case of conventional exf trusion. In some cases reduction in specific power consumption may reach 20 % [36, 39].

Thus, extrusion under conditions of combined shear produced by rotation or vibration of molding elements can be expedient and promising for the solution of one or several of the following technological problems: reduction of load on thrust units of the screws in extrusion machines (especially in processing of high-viscous polymers including the high-filled ones), improvement of the quality of surfaces of manufactured items, increasing the stability of dimensions of manufactured items due to reduced swelling of extrudate, reduction of process temperature, improvement of mechanical properties of products, etc. The generation of combined shear in the molding zone in the processing of the so-called superhigh-molecular polyolefins, for example, polyethylene with $MM \geq 10^6$, is also promising. In this case the super-imposition of core vibrations, at a high cohesion strength of material, can, apparently, result in the polymer's breaking away from the wall of the molding tool while flow changes into sliding without the stick-slip effect [48, 49], which is one of the causes of unstable flow, and with notable reduction in the resistance of molding heads and power consumption for extrusion.

In *conclusion* we would like to note that rotary and vibration effects upon molten and dissolved polymers are gradually finding a wider application. In recent years a number of original investigations has been carried out not only in the sphere of polymer processing but also in other fields, for example, in that of effects upon structure formation of electrorheological suspensions [52], sealing of rotating shfts with polymer fluids [53, 54], etc.

At the same time, the development of heads and other devices with mechanical rotating and vibrating elements intended for commercial applications meets serious physical difficulties.

High-intensity vibration effect upon dissolved and molten polymers is relatively simple and can be obtained with the help of superimposition of high-frequency ultrasonic (US) vibrations. Periodic (cyclic) shear and volumetric strain of the medium is attained in this case without 'moving' elements and due to elastic oscillations of US-vibrators of various design which function simultaneously as moulding elements (nozzles, cores) of extrusion heads.

The use of high-frequency technology should be especially advantageous since the efficiency of vibration effects upon rheological properties of viscoelastic materials is determined (according to G. V. Vinogradov & disciples) by exeeding a certain critical value of the amplitude of strain velocity $\dot{\varepsilon}_0 = \varepsilon \cdot \omega$ (here ε is strain amplitude; ω is angular frequency) so that at high frequencies a radical drop of the value of effective viscosity η can be expected.

3 Effect of Ultrasonic Vibrations upon Flowable Polymers

Usually acoustic cavitation (AC) is observed and studied in low-viscous fluids far from boiling mainly these are water, organic solvents and weak solutions of polymers. Fluids with a higher viscosity 'suppress' acoustic cavitation, brake the motion of bubble walls and eliminate thus the secondary cavitation effects having various technological applications: dispersion of solid particles in a fluid, interemulsification of two fluids, cleaning of polluted surfaces, depolymerization of weak polymer solutions, etc. For this reason the cavitation of high-viscous fluids has not been thoroughly investigated and no interesting effects have been revealed in this area until recently.

However, such viscous fluids as molten polymers can be very interesting objects for study from the point of view of ultrasonic (US) effects since they are characterized not only by viscosity but also by elasticity which alters radically their reaction to US-effects.

In the second part of this review we make an attempt to systematize the results of investigations into ultrasonic effects upon molten polymers carried out recently, and to prognosticate potential spheres of application for phenomena revealed therein. Note that this sphere has not been reviewed yet in literature, the occurrence of original publications is very low and the field of research is relatively narrow.

3.1 Brief Description of Acoustic Cavitation

3.1.1 Dissolved Polymers

Acoustic cavitation (AC), formation of pulsating cavities in a fluid, occurs when a powerful ultrasound is applied to a non-viscous fluid. The cavities are formed when the variable acoustic pressure in the rarefaction phase exceeds the cohesive strength of the fluid. Under acoustic treatment (AT), cavities grow to resonance dimensions conditioned by frequency, amplitude of oscillations, stiffness properties and external conditions, and start to pulsate synchronously (self-consistently) with acoustic pressure in the medium. The cavities undergo significant strains (compared to their dimensions) and their size decreases under compression up to collapsing. This non-linear behavior determines the active, destructional character of the cavities near which significant shear velocities, local pressure and temperature bursts occur in the fluid. Cavitation determines the specific character of acoustic treatment of the fluid and effects upon objects resident in the fluid, as well as all consequences of these effects.

Theoretical analysis of acoustic cavitation is focused, normally, on the study with the help of Nolting-Nepairas, Hering-Flinn equations and Kirkwood-Bete pulsations in the fluid of a single cavitation bubble [55, 56]. This approach features a number of principal drawbacks since it does not allow the explaination and quantitative assessment of such experimental facts as the drop of acoustic resistance, fluid radiation at cavitation, its heating related to the dissipation of primary acoustic energy and the presence of acoustic wave after the area of cavitation [57].

More adequate patterns of acoustic cavitation taking into account the motion of a cavitation bubble ensample are given, for example, in [58-60]. In [61,62] the authors consider a physical model of acoustic cavitation without the a priori non-linearity of the continuous medium, a mixture of fluid with similar precavitation bubbles pulsating linearly therein. It is assumed that the presence of 'radial' instability (tendency to collapse) of the bubbles results in formation of a flat stationary-form wave with a steep front on which the density of the equivalent medium grows by a leap (bubble collapsing wave) in the above-mentioned equivalent (model) medium in the phase of its compression. Similar representations are used in the case of loose or porous media having hollows. The steep wave front is formed in such media only due to peculiarities of the compression process itself, i.e., due to the property of these media of radically changing (by leap) the specific volume of pores (cavities) under load [63]. The above representations allow the writing down of the connections between parameters of the equivalent medium on two sides of the bubble collapse wave front with the help of conventional methods [64] and without the analysis of highly complicated processes of bubble collapsing inside the front itself.

In [62] the author has derived analytically, on the basis of the cavitation pattern under consideration, simple expressions for calculations of acoustic energy introduced by the vibrator into the fluid at cavitation and acoustic resistance of the fluid's radiation. These expressions have been also checked experimentally.

Viscosity of dissolved polymers drops irreversibly under acoustic treatment [65-68]. The depolymerization process us rather fast during the first minutes of the treatment and then it becomes slow and ceases completely when the equilibrium molecular mass (MM) M is reached. The higher the polymer's initial molecular mass N_0, the higher the rate of destruction. The majority of authors associate polymer destruction in solution with cavitation effects occurring under acoustic treatment.

It has been established that the degree in the limit of destruction alters depending on the power of applied acoustic energy and duration of acoustic treatment. Destruction of polymers is also affected by the nature of gases dissolved in the fluid. Increase in the concentration of the initial solution of polymer decreases the rate of destruction; destruction increases with growing external static pressure and then passes a wide maximum (the plateau) and drops.

Temperature dependency of the rate of destruction is determined by the effect of temperature upon the cavitation process there is an optimum temperature at which the intensity of cavitation is maximum; similarly, the rate and depth of destruction depend upon the frequency of acoustic vibrations.

Destruction of dissolved polymers under acoustic treatment results in the formation of macroradicals. Insonification of dissolved polymers in the presence of monomers or a mixture of polymers results in their polymerization; production of copolymers is also possible. In the presence of oxygen, radicals may combine following a peroxide pattern.

The efficiency of acoustic treatment drops and may produce no results at all at a significant rise of external pressure, viscosity of the solution or its temperature since the mechanochemical effects in acoustic treatment of dissolved polymers are determined, primarily, by the presence of cavitation and its intensity.

2.1.2 Molten Polymers

One of the first systematic description of acoustic treatment effect upon molten thermoplastics has been given, in [69]. Molten polysterene (PS), polyethylene (PE) and polyvinylchloride (PVC) were insonified, respectively, at temperature of 270 °C, 250 °C and 250 °C, frequency of acoustic treatment was 0.35, 1.0 and 3.0 MHz, and intensity was 3 to 5 W/cm^2. Ultrasound was applied to a polymer sample 3 mm thick through oil and a layer of foil; the sample was placed in an inert atmosphere. Molecular mass, heating, orientation, dispersion of pigment, and viscosity of the polymers under insonification was investigated. The absence of destruction was revealed under all conditions of acoustic treatment (duration 5 min.). Maximum heat release was observed at a frequency of vibration of 1 MHz, the minimum was observed at 3 MHz; Plasticized PVC was heated less than the other materials. During 10 minutes of acoustic treatment the temperature of material rose first, and then could even drop; the effect of heat exchange was not taken into account in this case. The authors of [69] discovered a certain orientation of macromolecules in the insonified PVC from anisotropy of its dielectric properties. Destruction pigment sinters in molten PE under the effect of acoustic treatment was also observed. No post-effect of acoustic treatment upon the viscosity of a polymer was revealed, measurements during insonification registered a decrease in effective viscosity by 5–10% but the authors took into account the heating of the polymeric material.

Investigations into the potential applications of acoustic treatment as a mechanical effect, from the point of view of processing of plastics, are aimed at intensification of manufacturing processes; another objective is to obtain products with improved properties. These problems can be solved both separately and simultaneously. To intensify the manufacturing process, the effect of thixotropic viscosity reduction in polymeric systems at periodic shear can be used [41, 70 – 73]. Investigations into the effect of shear vibration of infrasonic and sonic (up to 400 Hz) frequency upon viscoelastic materials have demonstrated that there is a critical amplitude of strain velocity the exceeding of which brings the effect into the non-linear are when rheological characteristics of the system begin to depend upon the amplitude of strains. These vibrations destroy thixotropically (reversibly) the structure of material. After the effect ceases, the structure is restored in a time measurable by minutes or hours; at first connections with short relaxation time are restored and then come less strong connections with a longer characteristic time. In filled systems the increased concentration of filler results in a radical decrease in the linear strain area and efficiency of ultrasonic vibrations grows. Destruction of structure was also observed in pure polymers but at high amplitudes of shear rate [71].

The amplitude of shear velocity is a determinant of the degree of thixotropic destruction of a material's structure, independently of the frequency of vibration effect. The effect of reversible viscosity drop is also observed at ultrasonic frequencies. Vibrothixotropy of concentrated polyisobutylene (PIB) solution was sMudied earlier at a frequency of acoustic treatment of 18 kHz and an amplitude of up to 15 mcm [74]. The first difference of normal stresses determined by the degree of extrudate's swelling drops proportionally to shear stresses. The exponent, in the so-called power-function flow principle, approaches, with increasing amplitude of acoustic treatment, 1 (Newtonian flow). Similar results have been obtained for molten thermoplastics: high-density polyethylene, shock-resistant polystyrene (SRPS) and PVC [75] with the help of a plant based on constant-pressure capillary viscosimeter with vibrating capillary (frequency of vibration was approx. ≈ 18 kHz, amplitude of displacement was between 8 and 20 mcm). The lower the velocity of stationary flow, the higher the efficiency of acoustic treatment, in the same way as in case of the low-frequency shear. In [76] a possibility of intensifying the extrusion of polypropylene with the help of acoustic treatment was studied. It has been revealed that acoustic vibrations result in a thixotropic reduction in the viscosity of molten polypropylene and a decrease the required molding pressure during extrusion. Power demand aspects (specific power consumption) of extrusion under acoustic treatment have been also considered.

In [77] the authors give dependencies of the maximum Newtonian viscosity upon amplitude of periodic strain velocity $\eta_0 = f(\dot{\varepsilon})$ for polyethylene and polystyrene. It has been also revealed that the dependency of normalized viscosity upon the velocity of stationary shear $\eta/\eta_0 = f(\dot{\gamma} \cdot \eta_0)$ obtained at $\dot{\varepsilon} = 0$ coincides with a similar dependency when acoustic treatment is employed, i.e., at $\dot{\varepsilon} \neq 0$. In other words, the effect of shear vibrations and velocity of stationary shear upon value η can be divided, representing the role of the first factor in form of dependency $\eta_0(\dot{\varepsilon}_0)$ and that of the second in form of dependency (η/η_0) upon $(\dot{\gamma} \cdot \eta_0)$ invariant in relation to $\dot{\varepsilon}$.

Figures 16 and 17 indicate that a significant reduction in effective viscosity of molten polymers is attained under acoustic treatment and consequently, there is a possibility

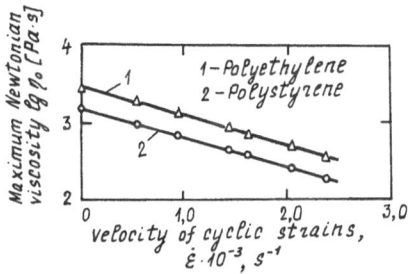

Fig. 16. Maximum viscosity versus periodic strain velocity $\dot{\varepsilon}$

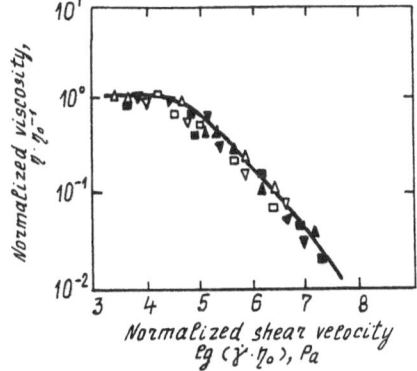

Fig. 17. Invariant characteristic of viscosity. Polyethylene ($\blacktriangle$ — $\dot{\varepsilon} = 9\,\text{s}^{-1}$; $\blacksquare$ — $\dot{\varepsilon} = 1.6 \times 10^3\,\text{s}^{-1}$; $\blacktriangle$ — $\dot{\varepsilon} = 2.3 \times 10^3\,\text{s}^{-1}$). Polystyrene ($\triangle$ — $\dot{\varepsilon} = 9 \times 10^2\,\text{s}^{-1}$; $\square$ — $\dot{\varepsilon} = 1.6 \times 10^3\,\text{s}^{-1}$; $\triangle$ — $\dot{\varepsilon} = 2.3 \times 10^3\,\text{s}^{-1}$)

of increasing the velocity of the material's flow under limited pressure in extrusion heads. At a constant flow rate the acoustic treatment allows a reduction of pressure in the head, i.e., the load on the thrust-bearing assembly of the machine's screw, a lowering of the extrusion temperature in order to reduce the danger of undesirable thermodestruction of polymers.

The possibility of increasing critical shear velocities $\dot{\gamma}_{cr}$ under superimposed vibration is conditioned, in principle, by the fact that relaxation properties of melts alter under acoustic treatment in the direction of reduced characteristic relaxation times θ_r. Proceeding from the known unstable flow criterion (elastic turbulence) $(\dot{\gamma}_{cr}\theta_r) = \text{const}$, the role of vibrations is limited to the reduction of η_0 and θ_r and, consequently, they allow the increase of $\dot{\gamma}_{cr}$ (or τ_{cr}). Respective experiments [77] carried out with different polyolefins reocaled a *symbateness* of dependencies of $\varphi = 1/\eta_0$ and $\dot{\gamma}_{cr}$ upon value $\dot{\epsilon}_0$. This corroborates the general explanation of the effect of vibrations upon critical conditions of deformation as a consequence of alteration of the relaxation properties of melts and the general concept of the causes of unstable flow as a consequence of relaxation transition in a polymer.

In [77] the authors also give an example of an extrusion head design with acoustic treatment of molten polymer in a circular molding clearance.

It has been revealed comparatively long ago that ultrasound, besides the thixotropic effect, reduces the friction coefficient and, therefore, acoustic treatment may also reduce the resistance of molding channels due to near-wall sliding of the material. Thus, for example, in [78] friction of rubber against metal was studied and it has been revealed that friction is reduced by a factor of 2 under insonification of a metal wall. Earlier works [79-81] deal with intensification of extrusion and casting of rubber mixtures with the help of acoustic treatment. Insonification of the molding channel of an extruder allows an increase in the rate of extrusion by a factor of 1.5–2 without a deterioration in the quality of extrudate's surface and a reduction in power consumption by the screw extruder by 20 to 30%. Acoustic treatment improves the quality of extrudate surface under extrusion at a rate of higher than 1.5 m/min. The maximum effect of acoustic treatment is observed in molding channels shaped as narrow slits in the processing of high-adhesion mixtures and "stiff" mixture compositions.

In [82] the authors describe the behavior of a viscoelastic fluid on the surface of an acoustic vibrator. The diagram shows that the fluid located on a horizontal surface of acoustic radiator, in the area of viscoelasticity, is acted upon by forces normal for this surface; the fluid swells above the radiator, takes a shape close to a spheric drop, then a thin neck is formed through which the fluid flows into the drop until it sinks to the surface under the action of gravity; then this process is repeated. These phenomena had not been described earlier in literature.

In [83] the authors studied the flow of polypropylene and polyethylene with the help of a constant-velocity viscosimeter developed on the basis of a standard extrusion plastometer for measurement of the melt flow index (MFI) to the stem of which acoustic vibration was applied.

In the processing of plastic masses the destruction is traditionally considered as a negative factor deteriorating physical and mechanical properties of products and manufacturers try to avoid it in every possible way. Mechanical destruction of molten polymers takes place, primarily, under the action of shear strains effectuating the tension of macromolecules [65,66]; in this case, molecules with a high molecular mass are the first to be destroyed and, in contrast to the cases of thermal and chemical effects, a monomer is not usually formed. The degree of destruction depends upon shear stress and the duration of its action while the molecular mass reaches a contsant value (M) with time and this value corresponds to the applied stress. Thus, there is a critical shear stress below which the destruction does not occur for each molecular mass. At higher temperatures the mechanical destruction can be accompanied by thermo-oxidizing reactions (thermomechanical destruction). As has been demonstrated in the known works by R. Porter et al., the intensity of destruction under mechanical effect passes its minimum, as temperature rises, since the mechanical destruction drops with increased mobility of macromolecules, but then the thermal destruction doped by the shear stress starts to develop [86].

High pressure is also a factor causing mechanochemical transformations in a polymer. Combined pressure and shear effect turned out to be most efficient (see, for example, reviews of this matter in [87-90]. Cros polyethylene containing up to 97% of gel fraction has been obtained under the action of super-high pressure of the order of 10^9 Pa in the presence of organic peroxides [87].

The analysis of the causes of destruction of molten polymers under acoustic treatment is very important from the theoretical and practical point of view. Thermal destruction can not be a determining factor if there is no significant heating of the material under insonification. The mechanical destruction caused by mutual friction and tension of macromolecules under acoustic treatment is also excluded since the dimensions of polymer molecules are by several orders of magnitude less than the acoustic wavelengths (at a frequency of 20 kHz the ultrasound wavelength in a molten polymer equals approximately 6 cm). Similarly to the acoustic treatment of dilute polymer solutions, an explanation proceeding from the cavitation mechanism of destruction suggests itself. However, the calculated collapse velocity of a cavity in a fluid with a viscosity above 1–2 Pa s is low and the secondary strong-acting effects of acoustic cavitation should not develop, theoretically, in such a medium [66,91].

Dynamics of isolated cavities in a viscoelastic fluid has been treated in several works. In this case, the relaxation of strain velocities is taken into account, in contrast to the case of a Newtonian fluid. In [92] the authors have considered some aspects of cavity dynamics in a viscoplastic medium. An earlier work [93] studied the dynamics of a cavity in a fluid with integral rheological equation of state of hereditary type. Pulsation of a bubble in a fluid described by the three-constant Oldroyde model has been considered in [94]. In [95] the authors carried out a numerical investigation into the equation of motion of a cavity's boundary in a fluid described by Reiner-Rivlin model equation. An accurate solution of the equation of minor oscillations of a spherial cavity in a medium described by the eight-constant Oldroyde model has been obtained in [96]. In [97] the effect of non-Newtonian properties of the medium upon the velocity of bubble boundary motion was considered. A numerical analysis has revealed that the fundamental frequency of bubble pulsations in a viscoelastic medium decreases significantly with the increase in viscosity and relaxation time of the medium; maximum bubble collapse velocities drop radically, suppressing thereby the secondary cavitation effects. It has been also pointed out in [94] that elasticity may reduce the viscosity's damping effect upon the process of bubble collapsing.

At the same time, it has been shown in [98] that the tensile stresses at which cohesive destruction of the material takes place do not exceed 1 MPa, in the case of flowable polyisobutylene with a Newtonian viscosity of 10^6 Pa s strained under conditions of the triaxial stressed state. This corresponds to static strength values of low-viscous fluids determined experimentally long ago [99].

Consequently, acoustic cavitation can also be expected in molten polymers under certain conditions at a relatively low intensity of acoustic treatment. High-viscous polymer systems characterized by elasticity or, in other words, demonstrating the properties both of liquids and elastic bodies simultaneously are a matter of special interest for a study of the behavior of materials in an acoustic field.

3.2 Acoustic Cavitation in molten Thermoplastics

In [100] we have found acoustic cavitation (AC) and some associate phenomena in flowable high polymers. Experiments have been carried out with the help of a constantpressure viscosimeter with round capillary. An ultrasonic vibrator was located in a viscosimetric tank at a certain distance from the inlet to the capillary. The study was conducted on high-density polyethylene. It has been established that the resistance to the polymer's flow through the measuring capillary is reduced significantly (up to 80 times) under conditions of acoustic treatment. We measured the spectral composition of the variable-pressure signal recorded by a miniature microphone in molten thermoplastic. It has been known for a long time [101] that the occurrence of multiple harmonic components in the spectrum of this signal indicates the presence of nonlinearly pulsating cavities in the fluid. The spectrum of the variable-pressure signal contained up to 4 to 5 multiple harmonics. Study of molecular-mass distribution in high-density polyethylene samples has revealed that the acoustic treatment causes mechanical destruction of the polymer. The authors have also carried out a visualization of the processes which take place under acoustic treatment of a flowable high polymer. Figure 18 gives a photo of polymer flow with cavitation bubbles. Experiments were conducted in a transparent capillary with flat-slit molding channel. It is evident that the 'cloud' of bubbles is formed at the corners of the acoustic radiator

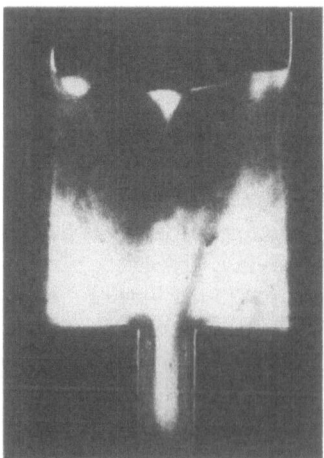

Fig. 18. Polymer flow with cavitation bubbles under acoustic treatment

and moves further with the polymer flow. Distribution of bubbles in the flat-slit channel itself is of special interest: the zone adjacent to the central line of the slit and near-wall layers are always free.

In [102] the authors describe the design of an experimental plant for studies of acoustic cavitation in flowable high polymers with the help of optic methods; the plant employs a flat-slit transprent-wall capillary; acoustic treatment of a polymer was carried out at a frequency of 17.8 kHz, and amplitude of vibration between 0 and 15 mcm. The study was conducted on 1,2 polybutadienes of narrow molecular-mass distribution; tests were arranged at room temperature. It has been demonstrated that static mechanical stresses occur in a stationary (non-flowing) polymer under the action of acoustic treatment; isochrome lines in the viscosimetric tank form a cellular structure with cell size of about 1–3 mm, and in the capillary the isochromes are observed in form of longitudinal strips (Fig. 19). The authors have also found that acoustic

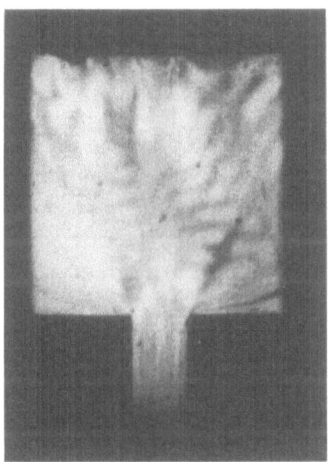

Fig 19. Isochrome lines in acoustic treatment of a stationary (non-flowing) polymer

treatment alters the pattern of polymer's stressed state at inlet to the capillary. Acoustic treatment of a polymer placed into a viscosimetric chamber results in the formation of areas of fine cavitation bubbles in the material; these bubles are observed in usual, non-polarized light. The above areas are formed near the vibrator face and walls of the chamber. They dissolve gradually after the acoustic treatment is terminated. We may assume that the formation of acoustic cavitation areas in a medium as viscous as the molten polymers has a character of destruction and is determined by the ability of a flowable polymer to respond to a high-frequency action similar to the reaction of a solid (quasi-solid) body.

It is known (see [91]) that acoustic cavitation occurs in low-viscous fluids (for example, in water) in a thershold pattern. In [102] the authors carried out an experimental investigations to estimate the thershold of acoustic cavitation in flowable high polymers. Figure 20 gives the results of experiments for three different samples. It can be seen that for one sample the critical (in terms of cavitation occurence) amplitude of acoustic vibrations depends upon time, i.e., in principle, it is of long-term strength

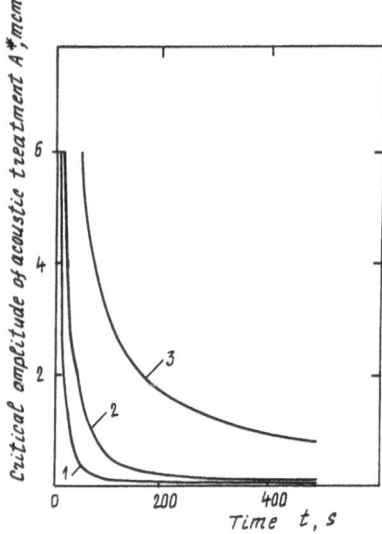

Fig. 20. Critical amplitude of acoustic treatment versus time

character. This similarity is quite logical since variable stresses occurring in the material when an acoustic wave passes through are determined by the amplitude of vibrations; occurrence of acoustic cavitation can be regarded as the moment of destruction (break in continuity) of the material. Relaxation times characteristic of the high-elastic and vitrous states of the polymer correspond to ultrasonic frequencies, so the disruption of the continuity of a molten polymer cab fillow patterns characteristic of solid (quasi-solid) bodies. This has been corroborated indirectly in [102] by mathematical processing of experimental data. Analytical dependencies of the induction period (time before the occurrence of acoustic cavitation) upon the amplitude of acoustic vibrations have been obtained in two forms: exponentional and power (see Table 1).

In this case the exponent describes better the branches of curves corresponding to high amplitudes, and the power function does the same for low amplitudes. In other words, at low amplitudes the material is destroyed in a way similar to that of a cross rubber while at high amplitude the pattern of destruction is similar to that of a solid body.

In [102, 103] we have also studied rheological characteristics of the flow of molten polymers and compositions under conditions of acoustic treatment. Experiments indicate that the flow-rate of polymers through channels can be increased significantly with the help of acoustic treatment of molten polymers. Figures 21 gives the relative increase in the flow rate of high-density polyethylene versus amplitude of acoustic treatment.

Increase in the flow of materials under acoustic treatment is conditioned by different factors. Investigation into rheological and molecular-mass characteristics of polymers having been subjected to acoustic treatment has revealed that, in the case of a low-intensity treatment, the effect is of a reversible (thixotropic) character. However, at high intensities of acoustic treatment, rheological characteristics of the material are not restored completely after the vibration effect is terminated and the

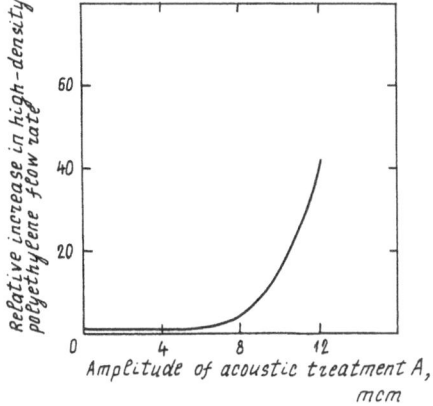

Fig. 21. Relative increase in high-density poly-ethylene flow rate versus amplitude of acoustic treatment

polymer's molecular mass is decreased. Acoustic treatment in high-filled compositions reduced the internal friction and facilitates the displacement of particles in relation to one another promoting thus the flow and mixing of the material.

Rheological and molecular-mass characteristics of polymer materials have been also studied after acoustic treatment.

Structural studies of high-density polyethylene samples subjected to acoustic treatment were carried out using the method of single-stage carbon-platinum replicas with the help of electron microscopy [104]. To identify the supramolecular structure, the authors employed the plasma pickling procedure. It has been demonstrated that microstructural elements of the material are comminuted under the action of acoustic treatment, the characteristic size of crystal formations diminishes. The large period changes insignificantly. The cross-size of lamellae suffers the maximum changes; the cross-size of crystallites practically does not change (Fig. 22). The character of crystallites practically does not change (Fig. 22). The character of crystallite aggregation

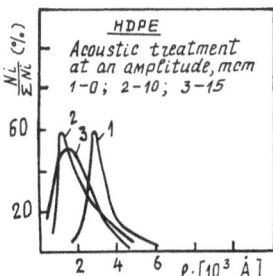

Fig. 22. Lamella cross-sectional dimension distribution curves

resultant from acoustic treatment changes; the degree of disorientation of crystallites increases with intensity of acoustic treatment and high-defective ('loose') areas take positions between them. In other words, the structure of samples after acoustic treatment is more defective in terms of all parameters, the stage of a polymer's transition from a melt with the maximum chaotic state to a textured system seems to be fixed therein. It is assumed that the changes observed in the structure of crystallized samples are related to the effect of acoustic treatment upon the process of formation, distribution and growth kinetics of the polymer crystallization embryos in visco-flowable state.

3.3 Practical Utilization of Acoustic Treatment

It follows from the review of above-mentioned works that the acoustic treatment of molten polymers and compositions improves significantly the flow of these materials. The effects of the acoustic field have various manifestations: from alteration of the stressed state of a flowing melt to the destruction of the polymer. Acoustic treatment of a relatively low intensity can be used to reduce resistance of molding tools to the flow of molten thermoplastics [76] and rubbers [105] with intensification of their processing. As a result of increased intensity of the acoustic treatment, acoustic cavitation occurs in a molten polymer causing irreversible changes in the material. An intensive acoustic treatment engaged in mechanical and chemical processes in molten polymers offers ample opportunities for a radical change in the properties of polymer materials.

Data on various mechanochemical transformations of polymers resultant from acoustic treatment of their dilute solutions under conditions of cavitation have been discussed in detail in review [66]. These transformations include the destruction of macromolecules, grafting of monomers in polymers and formation of block and grafted copolymers. All these processes have not found a wide practical application so far because the reaction medium is 'inconvenient' in terms of technology. Major methods of thermoplastic processing are based on their transition to the viscoflow state and therefore, the possibility of obtaining free radicals under acoustic treatment of *molten polymers* opens the way for mechanical and chemical transformations of thermoplastics directly at the stage of their processing into products. Compared to the known methods of mechanochemical destruction of macromolecules in a molten polymer (screw plastication, etc.), the acoustic cavitations offers wider opportunities to control the depth of the process. An important point is that standard processing equipment can be equipped with the acoustic treatment system. Multiple aspects of practical application of acoustic treatment of various materials and its great potential contribute to the formation of a new field of physical chemistry, namely the sonic chemistry, or acoustic chemistry, which has already been described in [106].

4 References

1. Fridman ML, Peshkovsky SL (1979) Plasticheskiye massy 7: 29
2. Azovskaya VA, Safonova IL (1981) Himicheskaya promyshlennost za rubezhom 8(224): 45
3. Mytzyl VA, Smirnova TN (1983) Himicheskaya promyshlennost za rubezhom 1(241): 1
4. Grozdova GV, Smirnova TN (1986) Himicheskaya promysh. za rubezhom 3(279): 1
5. Fridman ML (1988) Development of equipment for mixing of molten polymers, ZINTIhim-heftemash, Moscow
6. Vinogradov GV et al. (1970) Inzhenerno-physichesky zhurnal 19: 377
7. Prokunin AN, Fridman ML, Vinogradov GV (1971), Mehanika polymerov 3: 497

8. Malkin A. Ya, Isayev AI, Vinogradov GV (1975) Mehanika polymerov 2: 306
9. Vinogradov GV, Fridman ML, Malkin AYa, Yarlikov BV (1970) Rheol. Acta 9: 323
10. Basov NI, Lubartovitch VA (1979) Vibromolding of polymers, Himiya, Leningrad
11. Myasnikov VP (1961) Zhurnal prikladnoy mehaniki i tehnicheskoy fisiky 5: 76
12. Gutkin AM (1961) Colloidny zhurnal 23: 20
13. Buchman YuA et al (1984) Thermal mass transfer 5(2): 17
14. Buchman YuA et al (1985) Inzh.-phys. zhurnal 49: 221
15. Schulman ZP, Zadvornyh VN, Litvinov AI (1987) Rheodynamics of nonlinearly viscoplastic fluids in circular channels with movable walls. Acad. of Sc. Bel. SSR, Minsk, Preprint 45: 51c
16. Schulman ZP (1975) Convective heat & mass transfer in rheologically complex fluids, Energiya, Moscow
17. Colleman BD, Noll WI (1959) J. Appl. Phys. 30: 1508
18. Fredrickson AG (1960) Chem. Engng. Sci. 11: 252
19. Tanner RI (1960) J. Mech. Engng. Sci. 2: 21
20. Savins IG, Wallick GS (1966) AIChE J. 12: 357
21. Bortnikov VG, Kuznetzov NV, Tyabin NV (1967) Plasticheskiye massy, 8: 49
22. Tyabin NV, Bortnikov VG, Vachagin KD (1968) Plasticheskiye massy 3: 351
23. Fridman ML (1970) Rheological properties and high-speed extrusion of polypropylene. Summary of the thesis, INHS, Ac. Sci. USSR, Moscow
24. Prokunin AN, Fridman ML, Vinogradov GV (1971) Mehanika polymerov 3: 497
25. Winter HH (1973) Rheol. Acta 12: 1
26. Winter HH (1975) Rheol. Acta, 14: 764
27. Colleman BD, Markwitz H, Noll WI (1966) Viscometri flow of non-Newtonian fluids. Springer Tracts of Nat. Philos, Berlin Heidelberg New York
28. Fridman ML, Peshkovsky SL, Vinogradov GV (1981) Polym. Engng. Sci. 21: 755
29. Fridman ML (1980) Regulation of rheological properties of thermoplastics and compositions based thereon in order to intensify molding processes. Summary of the thesis, In. Chem. Phys. Ac. Sci. USSR, Moscow
30. Fridman ML (1977) Technology of processing of crystall. polyolefins, Himiya, Moscow
31. Fridman ML, Isayev AI, Smyslova ES (1973) Study of extrusion flow under conditions of molding tool vibration. Proceedings of 18th All-Union Conf. on HMC, Kazan, p. 226
32. Litvinov VG (1982) Flow of non-linearly viscous fluid, Nauka, Moscow
33. Vinogradov GV, Malkin AYa (1980) Rheol. of Polymers, Mir, Moscow
34. Prokunin AN (1985) Some effects in flow of viscoelastic fluids. Summary of the thesis, IVS Ac. Aci. USSR, Leningrad
35. Chang Dee Han (1975) Rheol. in polym. processing, Academic, New York
    Translation into Russian edited by Vinogradov GV and Fridman ML (1979), Himiya, Moscow
36. Genender MM (1985) Extrusion of PVC-compositions through heads with rotating core, Summary of the thesis, Mendeleyev Inst. Chem. Technol., Moscow
37. Tadmar Z, Gogos CG (1979) Principles of pölym. processing, Wiley, New York Translation into Russian edited by R. V. Torner (1984), Himiya, Moscow
38. Volov BM (1986) Study and optimization of design and technology for manufacturing of pressure hoses of PVC-compositions. Summary of the thesis, Lomonosov Inst. Fine Chem. Tech., Moscow
39. Panov AK, Vachagin KD, Fridman ML (1975) Effect of periodic mechan. strains upon flow of molten polymers in prismatic channels, In: Rheology of polymer and dispersed systems and rheophysics Minsk, vol 1, p 123
40. Ziprin MG (1977) Mehanika Polymerov, 1: 127, 2: 294, 6: 1093
41. Feitelson LA, Yakobsons EE (1977) Mehanika Polymerov 6: 1075
42. Feitelson LA (1985) Vibrothixotropy in polymers. Summary of the thesis, IMKM, Ac. Sci. Latv. SSR, Riga
43. Fridman ML, Konyshev YuV, Ivankov DV (1971) Injection unit of molding machine. Auth. Certif. USSR No. 291803, Bull. No. 4
44. Fridman ML, Prut EV (1984) Uspehy Himiyi, Ac. Sci. USSR 53(2): 309
45. Lipatov YuS (1984) Colloid chemistry of polymers, Naukova Dumka, Kiev
46. Chang Dee Han (1981), Multiphase flow in polym. processing, Academic, New York
47. Kissin Yu. V, Fridman ML (1977) Mehanika Polymerov 1: 143

48. Leonov AI, Prokunin AN (1983) Rheol. Acta 22: 137
49. Leonov AI (1984) Rheol. Acta 23: 591
50. Kissin YuV, Fridman ML (1975) Zhurn. Priklad. Spectroscopy 24: 929
51. Casulli G, Klermont GR, Von Zigler A, Mena B (1987), 3d Annu. Meet. Polym. Process. Soc. Program and Abstract, April 7–10
52. Ragotner MM, Gorodkin RG, Bukovitch IV, Smolsky MB (1983) In: Prikladnaya Mehanika i Rheophysica, Ac. Sci. Bel. SSR, Minsk, p 75
53. Buchman YuA, Sysoyev VI, Prokunin AN (1987) Auth. Certif. USSR No. 1295117, Bull. No. 9
54. Sysoyev VI (1988) Study of shaft sealing by viscoelastic fluids. Summary of the thesis. IMSS, Ural Scient. Centre, Ac. Sci. USSR, Perm
55. Flinn G (1967) Physics of acoustic caviation in fluids. In: Meson U (ed) Physical acoustics. Mir, Moscow, p 7
56. Akulichev VA (1968) Pulsation of cavities In: Roseberg LD (ed) Powerful ultrasonic fields, Nauka, Moscow, p 131
57. Fukusima K, Sanssen D, Kikuchi E (1972) Sound field characteristics related to operation of ultrasonic converters. In: Kikuchi E (ed) Ultrasonic converters. Mir, Moscow, p 353
58. Rosenberg LD (1968) Cavitation area In: Rosenberg LD (ed) Powerful ultrasonic fields. Nauka, Moscow, p 131
59. Kogarko BS (1964) Doklady AN SSSR 155: 779
60. Boguslavsky YuYa (1967) Akustichesky Zhurnal 13: 538
61. Peshkovsky SL, Yakovlev AD (1976) Akustichesky Zhurn. 22: 422
62. Peshkovsky SL (1986) Cavitation of fluid in acoustic wave In: Physical and chemical effects upon manufacturing processes, Metallurgiya, Moscow, p 93
63. Herrman V (1976) Determining equations of compacting porous materials. In: Shapiro GS (ed) Problems of the Theory of Plasticity. Mir, Moscow, p 178
64. Zeldovitch YaB, Reiser YuP (1966) Physics of shock waves and high-temperature hydrodynamic phenomena, Nauka, Moscow
65. Baramboym, NK (1978) Mechanochemistry of high-molecular compounds, Himiya, Moscow
66. Kazale A, Porter RS (1983) Polymer reactions under effects of stresses. Translation from English, Himiya, Leningrad
67. Elpiner IE (1963) Ultrasound. Physico-chemical and biological effects. Physmatgiz, Moscow
68. Simionesku K, Oprea K (1970) Mechanochemistry of high-molecular compounds, Mir, Moscow
69. Bernhardt E (1954) Ind. & Engin. Chem. 3: 742
70. Ferry J (1963) Viscoelastic properties of polymers, Izdatinlit, Moscow
71. Vinogradov GV, Yanovsky YuG, Isayev AI (1970) Effect of vibrations upon polymers. In: Vinogradov GV (ed) Progress of polymer rheology. Himiya, Moscow, p 79
72. Ziprin MG, Feitelson LA (1972) Mehanika Polymerov, 4: 689
73. Feitelson LA, Yakobson EE (1977) Mehanika Polymerov 6: 1125
74. Peshkovsky SL, Generalov MB, Kaufman IN (1971) Mehanika Polymerov 6: 1097
    Peshkovsky SL (1971) Ultrasonic intensification of polymer material extrusion processes. Summary of the thesis, Inst. Chem. Mach. Build., Moscow
75. Karelin YuM, Mazurenko YuS, Peshkovsky SL (1973) Device for ultrasonic effect upon flow of molten polymers. In: Polymer equipment and processing of plastic masses, Tehnika, Kiev, p 31
76. Fridman ML, Gul VE (1975) Plasticheskiye Massy, 9: 27
77. Fridman ML, Peshkovsky SL, Vinogradov GV (1981) Polymer Eng. & Sci. 21: 755
78. Popov AV, Ischenko VG, Buryachenko AG (1970) Proizvodstvo Shin, Resinotehnicheskyh i Asbestotehnicheskyh Izdeliy, 6: 16
79. Askatelov AI (1972) Kauchuk i Resina, 3: 17
80. Popov AV (1968) Kauchuk i Resina 10: 49
81. Buryachenko AG (1969) Kauchuk i Resina 12: 18
82. Fridman ML, Peshkovsky SL, Popov VL (1978) Colloidny Zhurnal 4: 819
83. Ivanov AV, Bylalov YaM, Ismaylov TM (1975) Zavodskaya Laboratoriya, 6: 717
84. Abbas KB, Porter RS (1976) J. Appl. Polymer Sci. 20: 1289
85. Goetze KP, Porter RS (1971) J. Polymer Sci. 35: 189
86. Regel VR, Sluzker AI, Tomashevsky EE (1974) Kinetic nature of the strength of solid bodies, Nauka, Moscow

87. Engel T (1967) Mod. Plast. 1: 175; 1: 257
88. Zharov AA (1984) Uspehy Himyiy, Ac. Sci. USSR. 53: 236
89. Enikolopian NS (1984) Macromol. Chem. 8: 109
90. Larsen HA (1975) J. Phys. Chem. 61: 1643
91. Knapp R, Dailey J, Hammit F (1974) Cavitation, Mir, Moscow
92. Csang WJ, Csen HCh (1965) Phys. Fluids, 8: 758
93. Fogéer HS, Goddard (1970) Phys. Fluids, 3: 1135
94. Tanasawa J, Csang WJ (1970) J. Appl. Phys. 41: 4526
95. Avanesov AM, Avetysyan IA (1973) Izvestiya AN SSSR, Mehanika Zhidkosty i Gaza, 4: 170
96. Levitzky SP, Listrov AG (1974) PMTF, 1: 137
97. Avanesov AM, Avetysyan IA, Listrov AG (1976) Akustichesky Zhurnal 22: 812
98. Vinogradov GV, Elkin AJ, Sosin SE (1978) Polymer 19: 1458
99. Kornfeld MI (1951) Elasticity and strength of fluids, Moscow
100. Peshkovsky SL, Fridman ML, Brizitzky VI, Vinogradov GV, Tukachinsky AI (1981) Doklady AN SSSR, 258: 705
101. Zarembo LK, Krasilnikov VA (1966) Introduction to non-linear acoustics, Nauka, Moscow
102. Peshkovsky SL, Fridman ML, Tukachinsky AJ, Vinogradov GV, Eniklopian NS (1983) Polymer Composites 4: 126.
103. Fridman ML, Peshkovsky SL, Vinogradov GV (1980) In: Rheology 8-th Intern. Congr. Rheology Napoli vol 2 p 485
104. Fridman ML, Chalyh AE, Peshkovsky SL, Tukachinsky AI, Enikolopian NS (1983) Doklady AN SSSR, 273: 1169
105. Bylalov Ya M, Ismaylov TM, Ivanov AV, Lopchin VV (1976) Kauchuk i Resina 5: 33
106. Margulis MA (1984) Fundamentals of acoustic chemistry, Vyschaya Shkola, Moscow

Editor: M. L. Fridman
Received December 21, 1988

Fundamentals of Low-Pressure Moulding of Polymer Pastes (Plastisols) and Thermoplastic Materials

M. L. Fridman[1], A. Z. Petrosyan[2], V. S. Levin[3], E. Yu. Bormashenko[3]

The publication reviews the reological properties of the two basic types of polymer pastes (plastisols), i.e. those showing and those not showing the thyxotropic effect. A detailed study is provided of the theory of the plastisols' low-pressure injection moulding along with the solutions to the mould filling for diversely shaped and sized runner-cavity systems. Specifically, solutions are offered for plastisol mould filling with regard to particular effects: viscosity anomaly, non-isothermicity, structuring (gelatinizing) of the mass, pressure levelling in the cavity. The problem of frontal junction of the flows and their interdiffusion ("self-healing") is considered in connection with the method of filling one mould through two runners.

Of particular technological significance is the problem of optimizing the moulding process in terms of minimizing the mould filling time. Physical concepts and concrete ratios are cited for determining the minimal cycle of moulding for particular combinations of the geometric characteristics of the runner and mould, under different temperatures and other factors. The methods of the engineering calculation have been experimentally tested. The experimental data proved to agree with the modern theory.

---

[1] USSR Research Institute of Plastic Materials, Perovsky pr. 35, USSR-111112 Moscow.
[2] States Pedagogical Institute, USSR. (Arm. SSR)-377200 Kirovakan.
[3] NPO "Mashpriborplastic", Rustaveli str. 39 USSR (Ukr. SSR) Charkov.

Advances in Polymer Science 93
© Springer-Verlag Berlin Heidelberg 1990

1 Introduction. Polymer Pastes. Conventional Methods of Moulding

The term "polymer paste" generally means that we are dealing with composite systems comprising individual polymers or their mixtures which show relatively low viscosity and the properties of viscous media. In other words, "paste" is an empirically formed notion applicable to any viscous (dough-like) bodies irrespectively of their composition, structure and the purpose of designation [1]. Polymer pastes came into being mainly in view of the need for processing poorly soluble and thermally unstable polymers. Since pastes have relatively high flowability at rather low shear stresses and temperatures, a number of relatively thin-walled articles may be manufactured (films, coatings, various parts). The articles moulded from pastes are then subjected to gelatination (gel formation) upon heating, and the pastes are hardened over the entire volume of an article. A large number of paste formulations is known to date, though all of them are united by one common attribute: the pastes are prepared by mixing suitable polymers with some — predominantly organic — liquid, in which the polymers do not swell at room temperature, but do swell upon heating.

On an industrial scale, the most commonly used pastes are those based on polyvinyl chloride (PVC), the plasticators being the mentioned organic liquids. Pigments, fire-retardants, antistatic agents, as well as other ingredients usually added to PVC may also be introduced.

Pastes (or PVC plastisols) are used in manufacturing artificial leathers, haberdashery, fire hoses, toys, vessels, linings, clothing accessories, articles of technological and everyday usage, for inner finishing of vehicles, and for coatings onto metals, glass, and other materials [2,3]. The articles and coatings from polymer pastes, or plastisols, due to relatively good physical and chemical properties, ageing resistance, ability to attain saturated colours through dycing, and other advantages have been competing successfully with rubbers, thermoplastic polymers (e.g., polyolefins), and some other materials for as long as five decades. It should also be pointed out that moulding from plastisols requires relatively simple production equipment and tooling which makes plastisol processing highly profitable.

Processing of low-viscous systems is of utmost interest from a practical point of view, since it demands relatively low pressures and hence low energy consumption as compared, for instance, with compression moulding of thermoplastics and thermosets. On the other side, PVC has low thermal stability, and its processing is normally carried out within the narrow region between the temperature of melting and the temperature of thermal decomposition. The PVC melt shows rather high effective viscosity and is sensitive to thermo-mechanical loading. In contrast, the pastes, including PVC plastisols, may be moulded at low (down to room) temperatures and pressures where the danger of thermomechanical decomposition is practically excluded, and a low-power-consuming process is therefore achieved.

Up to now, moulding of plastisols was carried out by using conventional methods; some of these will be shortly described below.

Dipping is carried out by immersing a prototype or an article into a plastisol-containing bath followed by removal and heating the retained liquid up to 170 to 180 °C; an article or a prototype may also be heated up to 80–180 °C. This method is used in making gloves, droppers, bushings, linings, in obtaining corrosion-resistant and easily removable coatings on tools and spare parts.

Mould Casting may be of two modifications: pouring into open moulds and pouring followed by pouring out (the so-called "inverse dipping"). This method is applicable to plastisols of low and medium viscosity showing pseudo-plastic or close to Newtonian flow. In the case of free filling-in, the production line contains casting machine, oven where articles are hardened due to gelatination, and cooling bath. In the case of casting with pouring out, paste is placed into a preheated mould (up to 80–100 °C), held for some time sufficient for form-keeping or until a skin layer is formed; surplus paste is poured out, and the mould with the layer adhered is placed into the furnace for gelatination. The method is used in making vessels, high boots, and other hollow articles.

Spraying is applicable to plastisols with viscosities from 1000 down to 11 Pa · s $(10^4$ to 1.1×10^2 P), upon shear rate increase from 0.1 up to 150 s^{-1}. The process is performed by using pneumatic pumps with high compression ratio, e.g. 24:1, through an airless injector-gun. The method is used in obtaining protective coatings of motor car bodies and for noise and chemical insulating coatings. Another modification of the method utilizes spraying in an electric field of high voltage.

Spreading is the main method for making artificial leather. It comprises material spreading onto a moving fabric belt with a spreading knife or smearing roller [4].

Rotational casting is usually used in making hollow articles such as vessels, dummies, dolls, buoys, floate, etc. The weighed amount of plastisol is given into a metallic sealed mould which then is rotated in three mutually perpendicular directions upon heating in a chamber furnace. Upon gelatination the mould is cooled, stopped, opened, and the article is taken out.

Extrusion moulding is mainly used in making wire insulation and elastic profiles. For slow extrusion, pastes with viscositoes of about 15–18 Pa · s (150–180 P) are normally used; for fast extrusion (hear rates of 10^3–10^4 s^{-1}), plastisols with viscosities of about 20–25 Pa · s (200–250 P) are suitable. Extruding machines for pastes should have long screws with narrow thread (small depth of screw channel). The cylinder temperature is kept at about 150 °C, and the head (die) temperature at about 180 °C.

It may therefore be concluded that the numerous methods of plastisol preparation and processing include the common practice in processing both rubbers and polymers. It should be noted that the method of plastisol processing and appropriate industrial engineering are strongly dependent on the so-called lifetime of the polymeric composition, i.e. the time of spontaneous material solidification due to gelatination. Plastisols with lifetimes of 2–6 weeks are considered as technical grade. They are processed right where they are made mainly by spreading into films, coatings, and artificial leather. Pastes with high lifetimes (2–6 months) are called special or commercial grade. They may be transported over large distances and processed by any of the above-mentioned methods. Materials with optimal lifetimes are prepared from PVC with a moledular mass of $(1.5-1.8) \times 10^5$ by addition of 40–50% plasticizer (by mass) [1,5]. The procedures of paste preparation have been reviewed by Masenko [2].

The new method of plastisol processing called, **low-pressure moulding** has been developed during recent years. The procedure involves the filling of closed moulds through inlet channels (runners). This technique is in many details similar to the process of compression moulding of thermosets but essentially different in the following. First, the relatively low viscosity of plastisols provides a possibility of utilizing low (down to atmospheric) pressures, of avoiding the need of melting (plastification) of the material, which makes possible the use of more simple equipment and moulds made of cheap and even non-metallic materials. Second, an essential difference lies is the mechanism of material solidification: thermoplastics are solidified upon cooling in the forming instrument, where the polymer undergoes transition from the highly viscous into the highly elastic state and then into the solid state which is accompanied by the crystallization of the material (for crystallizing thermoplastics). But plastisols in contrary are solidified upon heating due to intense structuring (gelatination) of the compound.

The *gelatination mechanism* may be shortly represented as follows [1,5]. Upon temperature increase, plasticizer penetrates slowly into polymer particles which are enlarged in dimension. PVC sinters disintegrate into primary particles (depending on their strength, the disintegration may begin even at room temperature). At elevated temperature, say 80–100 °C, plastisol viscosity remarkably increases, the excess plasticizer vanishes, and the swollen polymer grains come into contact with one another. At this stage of pre-gelatination the material looks homogeneous, though articles obtained at this point (with uncompleted gelatination) show low quality (poor strength and elasticity, raised fragility). Gelatination may be considered as accomplished only when plasticizer is uniformly distributed over the entire PVC volume, and the plastisol becomes an homogeneous body. The swollen primary structural particles of PVC are almost completely "welded" (fused together) thus forming plasticized PVC. Gelatination may be characterized by the temperature at which the process of structuring is completed. In such a case, the finished articles have the best physical and mechanical parameters.

2 Quantitative Analysis of the Conventional Methods of Plastisol Moulding

Fundamentals of the various methods of plastisol technology have been considered in a number of papers, nevertheless the theoretical analysis and the construction of mathematical models have not yet been completed so far. The *dipping* process has been studied in more detail; cf. [2,6–10].

The approach to the quantitative analysis and mathematical modelling of the dipping process is based on the solution of the well-known problem of physico-chemical hydrodynamics of the thickness of liquid layers retained on the surface of a body removed from the liquid (see, e.g., [11,12]). Upon the assumption that the body (support, prototype, mould) is taken out of the plastisol liquid vertically, the general relationships may be written in the following form [2,7,11–14]:

$$\delta = K(\eta v/\varrho g)^{1/2} \quad \text{where} \quad (\eta v/\sigma) \gg 1 \tag{2.1a}$$

$$\delta = \frac{0.93\,(\eta v)^{2/3}}{(\varrho g)^{1/2}} \quad \text{where} \quad (\eta v/\sigma) \ll 1 \tag{2.1b}$$

and generally

$$\delta = \left(\frac{\eta v}{\varrho g}\right)^{1/2} f\left(\frac{\eta v}{\sigma}\right) \tag{2.2}$$

where δ is the thickness of the liquid layer σ the plastisol surface tension, v the linear velocity of the solid body removal from the plastisol, K a factor usually taken as equal to $^2/_3$ [12], η the viscosity of the liquid, g the free fall acceleration, and ϱ being the plastisol density.

Equations (1) and (2) hold true when the curvature radius, R_{curv}, of the body taken out of the liquid paste is much greater than the capillarity constant, i.e. when:

$$R_{curv} \gg \left(\frac{\sigma}{\varrho g}\right)^{1/2} \tag{2.3}$$

In some practically important cases, the function $f = \eta v/\sigma$ takes the following values [2,7,11–14]:

$f = \sqrt{2}$ if the back draining off is neglected;

$f = 1$ if the back draining off is taken into account for the case of the plate of infinitely large length;

$f = ^2/_3$ for the plate of finite length and for mean layer thickness.

The dimensionless criterion factor $f = \eta v/\sigma$ is of great importance in estimating liquid flow parameters; the discussion of its values is beyond the frames of the present work (as outlined in the references cited). It may be noted, however, that the same expression appears in the equation evaluating the ability of polymers for thread and fiber spinning [15]. To the first approximation, the criterion $f = \eta v/\sigma$ may be treated as a measure of shape-keeping ability of liquid in flow.

A lack of experimental data in the literature makes the check-up of Eqs. (2.1)–(2.3) rather difficult to perform, most of the data have been obtained for Newtonian liquids [12,14]. Equation (2.3) was shown to be valid at $f = \eta v/\sigma = 0.3255$; at $f > 2.4$ the formation of waves and "disturbances" of liquid flow was observed. The lower the viscosity of a system, the easier these phenomena are to be observed. The value of K was experimentally shown [14] to be equal to 0.657, i.e. rather close to the theoretically predicted value $K = ^2/_3$.

No essential difference was found under the experimental conditions between the layer thickness for Newtonian and non-Newtonian liquids, as may be judged from a few data available on non-Newtonian liquids (e.g., latex SKS-30L and dextrin glue). This non-trivial experimental result has not yet been convingly or reasonably explained.

For the process of continuous plastisol coating by dipping in the range of velocities of practical interest, Masenko and co-workers [2,7,16,17] suggested the following empirical formula:

$$\delta = \frac{\bar{\eta}_{lig}^{0,5} v^{0,45}}{(\varrho g)^{0,45}} \tag{2.4a}$$

or

$$\delta = c\left(\frac{\bar{\eta}_{lig} v}{\varrho g}\right)^{0,5} \tag{2.4b}$$

where $\bar{\eta}_{liq}$ is some effective (averaged) viscosity of a plastisol in the process of gelatination upon going from the initial state characterized by Newtonian viscosity to the final state close to complete loss of fluidity (when viscosity sharply increases); c is the conversion factor for velocity.

Such a simplified approach nevertheless proved to be practically acceptable in view of the fact that $\bar{\eta}_{liq}$ may be taken as constant for the plastisol of the same formulation and preparation time. This probably follows from the relatively low initial viscosity as compared to that aquired by the material upon heating.

If one takes into account the shrinkage of the fabric belt upon heating (the loss of liquid due to evaporation is neglected), and Eq. (2.4) may be transformed into [2,16]:

$$\delta = \frac{c}{(1 - \varepsilon_d)(1 - \varepsilon_l)} \left(\frac{\bar{\eta}_{lig} V}{\varrho g} \right)^{0,5} \tag{2.5}$$

where ε_d is the belt shrinkage in transverse direction (for circular textile jacket used in manufacturing fire hoses it is the shrinkage along the diameter); ε_l is longitudinal shrinkage. Both values are given in relative units as a fraction of unity.

Attempts to give a quantitative analysis of plastisol *extrusion* were undertaken only in a few published papers. They were based on the analysis of plastisol viscosity as a function of temperature and time. If in the processing of thermosetting plastics their viscosity is assumed as practically independent of time (except of materials sensitive to structural and chemical transformation in temperature and stress fields which are accompanied by thermo-mechanical decomposition and cross-linking of macromolecular chains, the extent of the larter being influenced by the time of exposure to thermal and mechanical loads [18-21]), then at extrusion of plastisols, in view of their gelatination, the additional condition should be satisfield:

$$\tau_{th} \geq \bar{\tau} > \tau_g \tag{2.6}$$

where τ_{th} is the time of plastisol thermal stability, $\bar{\tau}$ the mean material residence time in an extruder, and τ_g being gelatination time (structuring of material up to the complete loss of fluidity). In the opposite case, the plastisol will not be converted into melt, or the obtained product will have low physical and mechanical parameters.

According to Eq. (2.6), successful extrusion of polymer pastes is possible if the residence time and the intensity of mixing are varied independently. Most often this is achieved in screwless machines equipped by delivery proportioning pumps and a mixing shaft in a cylinder, the pump having an independently controlled drive [17, 22-24]. Redent advances (USA, FRG, Japan, etc.) in extruder design — enhanced mixing efficiency and maintaining the required temperature profiles by utilizing microprocessors and computing systems with feedback — open new horizons in improving the extrusion processing of polymeric materials, including plastisols. The equipment for preparing mixtures of thermosetting plastics and flowing systems involving these materials has been reviewed recently by one of the authors [25]. In our opinion, the novel camera-disc machines (called "discopack") suggested by Tadmor and utilized by Farrell Co. (USA) seem to be rather promising for mixing, pumping, and squeezing (extrusion moulding) of plastisols. A detailed description of the operation principle, the design of mono-, di-, and multidisc (multistage) discopacks, their quantitative analysis, the design techniques for high productivity and mixing efficiency is documented in patents [26-28] and reviews [25, 29, 30].

In plastisol extrusion, the cylinder is completely filled and low hydrodynamic resistance is attained. In view of this, the productivity of plastisol extrusion is much larger as compared to that of standard PVC processing. The correlation between the processing and design parameters of plastisol extrusion was derived by Masenko and co-workers [2, 17]. In the derivation, the expression for maximal shear rate in the circular clearance of the operating unit was substituted in the expression correlating the residence time of plastisol in an extruder to volume productivity. The authors took into account Eq. (2.6) and assumed that the inner diameter of a cylinder is much larger than the clearance between the shaft and the inner surface of a cylinder. The proposed correlation has the form:

$$\tau_g = \pi n \, D^3 \omega / 2 \dot{\gamma}_n Q \tag{2.7}$$

where ω is the rotation speed of a mixing shaft, $n = L/D$ is the dimenionless ratio of the cylinder length L to its inner diameter D and Q is productivity (output).

Spreading (smearing) is the most widely used and well-studied method of plastiso1 processing for making linoleum, artificial leathers, wire enameling, etc. Descriptions of the process and its quantitative analysis may be found elsewhere [3, 4, 31].

Other methods of plastisol processing (rotational moulding, inverse dipping, casting, continuous dipping) are described in a number a books, e.g. [1, 3, 4, 32], but their quantitative analysis is lacking. This also applies for the recently suggested method of low-pressure injection moulding. The gap apparently, however, has been closed by our recent by work [33-36], where a theoretical analysis of plastisol moulding was given, and the mathematical models and design technique were proposed for all the stages of moulding. These data were supported by experimental results. This will be the topic of the following sections. The presentation of the fundamentals of the low-pressure moulding of plastisols will be proceeded by a description of published and original data on rheological (viscosity) and technological properties of these materials.

3 Rheological and Processing Properties of Plastisols

3.1 General Rules. Influence of Various Factors

Plastisol properties depend on a number of factors, the most important among which are the nature of the polymer (PVC), the identity of plasticizer and other additives, the composition of formulations, the conditions of preparation and storage, the load-velocity (first of all, stress and shear rate), and temperature conditions of processing. Depending on these parameters, the character of plastisol flow may vary widely. That is why, in spite of a number of papers devoted to the study of the rheological (viscous) properties of plastisols (e.g. [2, 5, 37, 38]), these still can not be considered as well studied and widely accepted.

The rheological properties of plastisols generally could not be characterized on the basis of viscosity at some fixed shear rate (effective viscosity). Nevertheless, for practical purposes plastisols are conventionally classified into low-viscous (1–3 Pa · s), medium-viscous (10–15 Pa · s), and high-viscous (100–1000 Pa · s) plastisols. Viscosities are measured at low shear rate $\dot{\gamma} = 1\ s^{-1}$ when plastisol flow is Newtonian.

Plastisol properties as dispersed systems may be described by using some notions of physico-chemical mechanics; the following assumption are made:
— PVC particles are spherically shaped and uniformly distributed over the entire volume, the particles are small;
— there are no chemical reactions between PVC and plasticizer;
— the flow is laminar;
— the system viscosity is homogeneous, i.e. flow is Newtonian.

The viscosity of such an idealized plastisol may be evaluated [37, 39] by using the well-known Einstein formula:

$$\eta = \eta_0(1 + K\varphi) \tag{3.1}$$

where η and η_0 are the viscosities of plastisol and plasticizer, respectively; φ is the volume fraction of PVC powder with respect to the total volume of plastisol and K is the so-called shape factor for PVC particles (for spherical particles $K = 2.5$).

Having determined φ experimentally, one may roughly (to the first approximation) evaluate the viscosity of the plastisol. The real plastisols of course are not ideal dispersed systems, and their viscosity as a rule is essentially different from that given in Eq. (3.1). This results not only from the low accuracy of φ determinations and the deviation of particle shape from the ideal sphere ($K \neq 2.5$), but also from the strong influence of a number of factors which will be discussed below on the basis of the available published data.

The type of chosen polymer and additives most strongly influences the rheological and processing properties of plastisols. Plastisols are normally prepared from emulsion and suspension PVC which differ by their molecular masses (by the Fickentcher constant), dimensions and porosity of particles. Dimensions and shape of particles are important not only due to the well-known properties of dispersed systems (given by the formulas of Einstein, Mooney, Kronecker, etc.), but also due to the fact that these factors (in view of the small viscosity of plasticizer as a composite "matrix") influence strongly the sedimental stability of the system. The joint solution of the equations of sedimentation (precipitation) of particles by the action of gravity and of thermal motion according to Einstein and Smoluchowski leads [37, 39] to the expression for the radius of the particles, r, which can not be precipitated in the dispersed system of an ideal plastisol. This expression has the form:

$$r^5 = \frac{27RT\eta_0}{4\pi N(\varrho_n - \varrho_0)\ gt} \tag{3.2}$$

where R is the gas constant, N the Avogadro constant, T absolute temperature, g free fall acceleration, η_0 plasticizer viscosity, t time, ϱ_n and ϱ_0 are the densities of polymer and plasicizer, respectively.

At standard storage conditions, the critical value of r = 2.23×10^{-4} cm (or d = 2r $\approx 4.5 \times 10^{-4}$ cm) [37]. Equation (3.2) was obtained under the above — mentioned assumptions but it properly reflects the well-known fact that the plastisols of finely dispersed PVC are much more stable upon storage as compared to the pastes with large polymer grains. From Eq. (3.2) it follows also that the temperature increase will promote the sedimental stability of the material. However, it should be kept in mind that plastisol properties remarkably and quickly change at elevated temperatures, so that the sedimental stability of the system becomes of no importance, and the material viscosity becomes to be determined by the process of swelling (and subsequent dissolution) of PVC in plasticizer. It is taken for granted that the assumptions adopted in deriving Eqs. (3.1) and (3.2) cease to be fulfilled, and thus the formulas can not be used in calculations under these conditions [2,17].

The character of plastisol flow was received, at first, with controversial attitudes. Numerous authors [40-43] have outlined that plastisol flow is close to Newtonian flow within a wide range of tensions, τ, and shear rates, γ̇. Other authors state the pronounced abnormality in plastisol viscosity manifesting itself as either pseudo-plasticity [1,2,5,44-46] or dilatance [1,6,47,48] in various plastisol systems. Such contradictory statements, in our opinion, are not inherently inconsistent, since the data probably were obtained with plastisols of different composition and, which is more important, for different physical and chemical states of the samples. As one may assume, the freshly prepared samples which undergo no essential gelatination show Newtonian flow at low γ̇, whereas increased times and temperatures of mixing or storage, elevated T and γ̇ (or τ) at testing will result in some structuring, and finally in various abnormal phenomena upon variation of viscosity. The flow of plastisols will be discussed in more detail below.

Plasticizers also specifically influence plastisol properties. As a rule, they tend to lower the viscosity of a system proportional to their intrinsic viscosity [22,37,41]. Plasticizers are added in amounts from 20 to 150% of polymer mass, predominantly 20–80% (most frequently 35–55%) by mass. The lower limit is set by the minimal amount of plasticizer necessary for plastisol formation, the upper limit by its maximum amount which can ge bound by PVC [1,41,47].

Compatibility of PVC with plasticizer (Pl) also strongly influences the rheological and processing properties of plastisols [48]. The effects of the diverse plasticizers on PVC varies since some plasticizers (even well compatible with polymer), e.g. dibutylphthalate (DBP), may readily escape from the obtained materials and articles [17,23,43]. Plasticizers which are poorly compatible with PVC (the so-called "secondary Pl") may be used for the purpose-oriented change of viscosity properties of a system (fluidity increase) and saving the base Pl component.

The effect of various plasticizers was studied for a number of plastisols prepared both from emulsion and suspension PVC [2,6,7,37,42-46,49]. Judging from the published data [48], the most viscous plastisols are formed with mesamole, low-viscosity plastisols with dioctyladipate taken as a base, irrespectively of PVC type. Plastisol viscosity may be controlled by certain additives: small smounts of certain solvents may lower plastisol viscosity by as much as an order of magnitude [37,41], the use of bentonites makes pastes more dense [40,48]. Thermoplastic polyethylene may also be used as a thickening agent [6,42].

Temperature is a factor which strongly influences the viscosity of plastisols [2,22,

[45, 48)]: before gelatination, viscosity diminishes with temperature, this dependence obeying the well-known exponential law of Arrhenius (up to temperatures of about 40–60 °C). It should be noted that the temperature effects on viscosity were studied only qualitatively by all the authors in view of experimental difficulties [45], in most cases temperature rise was linear. Below we shall consider the simple and efficient experimental approach which eliminates experimental difficulties in studying $\eta = f(T)$ and which provides for obtaining reliable quantitative data. This approach was used in our recent studies [35, 36], since the lack of reliable data on the rheology of plastisols makes impossible the solution of theoretical and design problems.

When the temperature is raised according to the linear law, the function $\eta = f(T)$ predominantly shows extrema: at first it drops and then grows. The extremum at the initial stage of gelatination was found to result from the lowering of plasticizer viscosity, while the subsequent increase seemed to result from plasticizer penetration into polymer [7, 17, 23, 50]. This pattern of temperature dependence is observed at any shear rate $\dot{\gamma}$, which implies the predominant role of temperature at the initial stage of gelatination. Minimal viscosity is attained [2, 23, 45] at temperatures of 40–70 °C. Heating plastisols up to 30–40 °C and above in the course of preparation and storage leads to irreversiable changes of viscosity due to PVC swelling [5, 44, 49].

The effect of the *molecular mass* (MM) *of* PVC on the viscosity of pastes has been only slightly studied and unambiguous conclusions could hardly be drawn. Some authors [5, 49] report that the molecular mass of PVC (the Fickentcher constant) has practically no influence on η and the processing properties of plastisols (this conclusion of course can not be extended to the properties of finished articles). On the other side, subsequent communications (e.g. [51]) have reported on the essential influence of MM. Moreover, the following order of influence on the viscosity of a system was suggested [51]: MM (the Fickentcher constant), content of low-molecular fractions in PVC granulometric content of polymer, amount of emulsifying agent, morphology of PVC particles, etc. As is seen, MM is the first in this order, and plastisol viscosity was found to increase with increasing MM. In our opinion, the available experimental data are insufficient for such a definite conclusion to be drawn since the experiments [51] were conducted with PVC samples for which MM was varying within relatively narrow limits characterized by the Fickentcher constant between 60 and 68.8.

3.2 Rheological Properties of Industrial Plastisols as Applied to Injection Moulding (Experimental Data)

As follows from the preceeding section, plastisols represent thyxotropic media, the viscosity of which at $T \lesssim 40$–60 °C irreversibly increases with time due to gelatination. From the standpoint of their rheological properties, industrial plastisols may be classified into two types:
— with vaguely pronounced thyxotropic properties, i.e. weak abnormality of viscosity with practical independence of flow from break time (type I);
— with distinctly pronounced thyxotropy, i.e. strong abnormality of viscosity with pre-stationary motions dependent upon break time (type II).

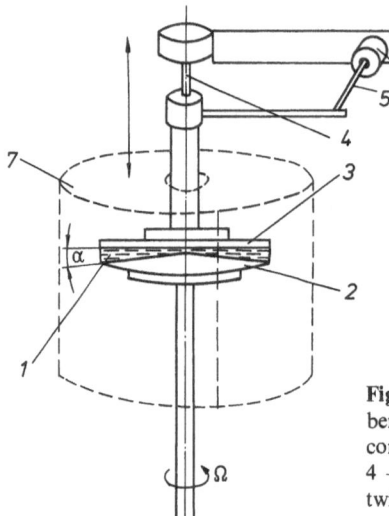

Fig. 1. Schematic layout of the operating unit of the Weissen-berg rheogoniometer: 1 — the liquid studied; 2 — rotating cone of the operating unit; 3 — round plate ("plane"); 4 — torsion beam; 5 — bar which displaces upon torsion twisting; 6 — induction translation transducer; 7 — opening camera with a thermostated furnace

For the purposes of injection moulding, it is of interest to consider the rheological properties of both materials aiming at elucidating their similarity and distinctions at modelling and designing the process of mould filling. *Plastisols of type I* (of weak thycotropy) may be exemplified by the composition of the following formu-lations (from Bata Co., Canada): PVC resin (mixture of Nipeon A-21 resin, 35.2%; Vinnik 70BX resin, 13.1%; Vinnik 37H resin, 2%), mixture of plasticizers 1249 in the amount 47.7% (butyloctylphthalate, 25%; di-2-ethylhexyladipate, 30%; centi-sizer 711, 40%; monoplex 5–73, 5%), stabilizer 704D, 1.5%, the rest being dyeing paste (cobalt naphthenate).

Plastisol of type II (of strong thyxotropy): PVC resin (Solwic), 54%; plasticizer (dioctylphthalate, DOP), 43%; stabilizer (DOPTC), 3%.

These materials were studied by the authors [35, 36] by using the Weißenberg rheo-gonimeter of Sangamo Co., England, with an operating unit of the "cone-plane" type (Fig. 1). In the experiments, the deformation rate $\dot{\gamma}$ and shear stress σ were calculated by using the following expressions:

$$\dot{\gamma} = \Omega/\alpha; \qquad \sigma = 3M/2\pi r^3 \qquad\qquad (3.3)$$

where Ω is the angular velocity of the operating cone rotation, α the angle between the cone and the plane of the operating unit, r the radius of the plane (3), M the torque at the torsion beam (4).

It should be noted that Eq. (3.3) is valid only at small $\alpha(\alpha \lesssim 3°)$ when the stress field in the liquid may be treated as constant.

The displacement of the bar (5) is measured, and the calculation is carried out by using [52]:

$$\sigma = \frac{3,82Kt}{8r^3} \Delta \qquad\qquad (3.4)$$

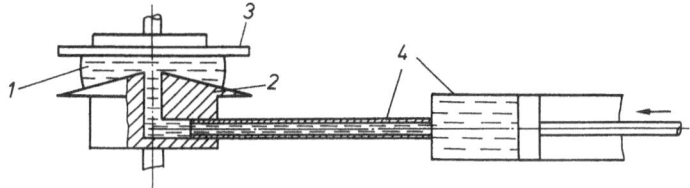

Fig. 2. Schematic layout for the modification of the operating cone in the rheogoniometer and the plastisol extrusion into the preheated operating cell: 1 — sample; 2 .— cone with drilling; 3 — operating plane; 4 — syringe with a feeding tubing

where Δ is the displacement of the bar (5) upon rotation of the operating unit and K_t is the torsion rigidity. In our experiments, $Kt^1 = 0.91 \times 10^2$ and $Kt^2 = 0.98 \times 10^3$ dyne $\cdot$ cm/μm, so that two torsion beams were used to check up the torsiometer rigidity calibrations. Also in the experiments, $\alpha = 2°$ and $r = 2.5$ cm.

It is also important to note that the conventional operating scheme of material loading into the clearance between the cone and plane (or into another operating unit, e.g. the reservoir of a capillary viscosimeter) and subsequent thermostatting is apparently unsuitable for studying the rheological properties of plastisols. Two possible approaches to the problem may be envisaged: to design a powerful heating system of the operating unit with the liquid under investigation [53], or to spray material into the operating unit preliminarily heated up to the given temperature. The latter procedure used by us seems to be more simple and less expensive (see Fig. 2). Liquid was sprayed through a small slit in the thermostatting furnace (7) with a syringe via tubing 4 mm in adiameter and 14-cm length in time up to 10 s. Plastisolwas sprayed through the 4.2-mm drilling in the operating cone, after which the thermostat was closed. Heating time for the sprayed plastisol may be evaluated as

$$t^* < \frac{h^2_{max}}{d} \approx 1 \text{ s (here } h_{max} \text{ is the largest clearance between the cone and plane and d}$$

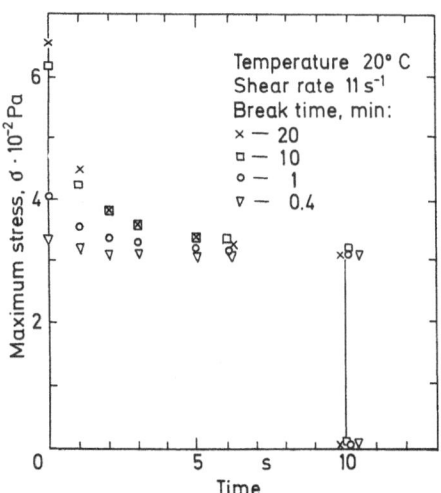

Fig. 3. Maximum stress σ_m (Pa) as a function of time t_s for plastisol II at $\dot{\gamma} = $ const; varied break times between experiments

is thermal diffusivity). During spraying time ($\leqslant 10$ s) the temperature of the operating unit remained practically unchanged. Temperature was maintained within ± 0.5 °C, experimental uncertainty for the plastisol of the same batch did not exceed 10%.

Studies on the *pre-stationary behaviour* of plastisols under conditions of shear in the absence of essential gelatination showed that plastisols of type I are similar to Newtonian liquids and do not display the so-called "characteristic time". In plastisols of type II the maximum stress is attained almost instantly (<0.1 s) upon fast application of the shear rate $\dot{\gamma} = $ const this, however, is followed by a decrease of stress (or effective viscosity $\eta = \sigma/\dot{\gamma}$) as a function of time, and the stationary flow is attained in several seconds (see Fig. 3). The instant cessation of motion results in a sharp drop of stress down to zero. This behaviour is in agreement with the published data [51].

The attained value of σ_{max} at $\dot{\gamma} = $ const (Fig. 3 presents the data for $\dot{\gamma} = 11$ s^{-1}) and the rise time of the stationary flow t_{st} are dependent upon the break time t_{br} between experiments: points 1–4 in Fig. 3 correspond to the break times of 1200, 600, 60, and 24 s, respectively. Upon the break time increase, the values of σ_{max} and t_{st} also increase; however, at times periods of longer than 15 min (experiments were performed up to 1 h) they become independent of time. The behaviour is qualitatively the same at fixed break times and temperatures but for various $\dot{\gamma}$. Moreover, the ratio of σ to stationary stress σ_{st} as a function of time is independent of deformation rate (see Fig. 4). For the same t_{br} and different T the functions $\sigma(t)$ differ one from another by a constant number. Therefore, temperature (just as $\dot{\gamma}$) has no influence on the ratio σ/σ_{st} as a function of t, which implies that the influence of T on σ may be qualitatively studied in stationary flows.

The question arises whether the studied thyxtropic phenomena should be taken into account in mould filling with plastisols of Type II, i.e. to assume medium instantly reacting to loading like plastisols I. This is supported by experimental data used in Fig. 4. As is seen, the stress responds instantly both on its fall and rise. Hence,

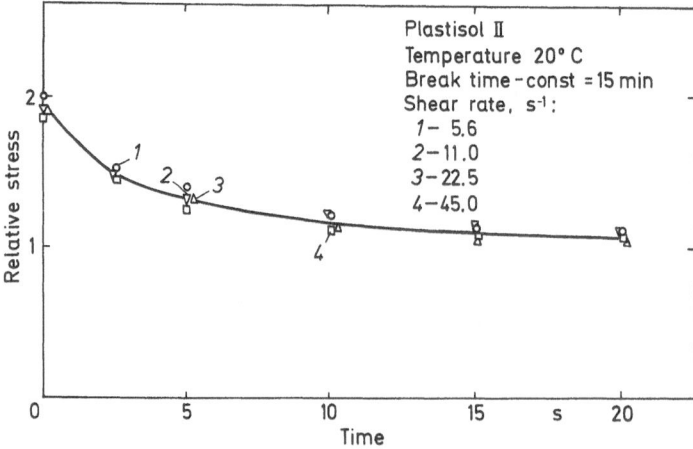

Fig. 4. Relative stress σ/σ_{st} as a function of time t for plastisol II at T = 20 °C, fixed break time between experiments (t_{ret} = 15 min), and varied shear rates

the thyxotropic phenomena even more so may be neglected in mould filling where the deformation rate is changed rather smoothly. This conclusion is also supported by the fact that the plastisols used in injection moulding have the so-called "damaged" structure (premixed in an apparatus with a stirrer), while the relaxation time, which is necessary for stress to decrease essentially even in "relaxed" plastisols, has a value of about 1 s (cf. Fig. 3); this is by an order of magnitude lower than the time of mould filling in real processing.

Stationary flow of plastisols by the action of shear (which may be evaluated, for example, from the fact of stress constancy during 30 min [36, 54]) may be attained only at certain limited temperatures. In our experiments [36, 54, 55] this was observed at $T \lesssim 60 °C$ for plastisols I, and at $T \lesssim 45 °C$ for plastisols II; at higher temperatures gelatination set in.

The available experimental data for plastisols of both types [34–36, 55, 54] lead to the following important conclusions. Plastisols of type I show a slight abnormality of viscosity: viscosity $\eta = \sigma/\dot{\gamma}$ becomes practically independent of $\dot{\gamma}$ and equal to the lowest Newtonian viscosity $\eta_\infty = 3.5$ Pa · s at 21 °C and $\dot{\gamma} \gtrsim 10^{-1}$ s^{-1}. For type-II plastisols the abnormal behaviour is more pronounced: at large $\dot{\gamma} \approx 5 \times 10^2$ s^{-1} the lowest Newtonian viscosity $\eta_\infty = 5$ Pa · s is attained at 20 °C, the effective viscosity depends of $\dot{\gamma}$ (according to a power law); at low rates $\dot{\gamma} < 10^{-1}$ s^{-1}, σ becomes to be slightly dependent on $\dot{\gamma}$ (for example, at 20 °C $\sigma = \eta_{eff}\dot{\gamma} \approx 50$ Pa), i.e. the system may be considered as posessing yield strength (like the Bingham-Shvedov body).

For both plastisols, the functions $\sigma(\dot{\gamma})$ [or $\eta_{eff}(\dot{\gamma})$] obtained at different temperatures are linearly dependent. So, for plastisol I, for instance, the curves obtained at 40 and 56 °C may be obtained from the dependence $\sigma(\dot{\gamma})$ at 21 °C by mere multiplying by 0.19 and 0.43, respectively; for plastisol II, the function $\eta(\dot{\gamma})$ at 30 and 43 °C may be obtained from that at 20 °C by multiplying by 0.58 and 0.76, respectively.

Therefore, the effective viscosity of plastisols may be written as follows:

$$\eta_{eff}(T, \dot{\gamma}) = \eta_\infty(T) \cdot f(\dot{\gamma}) \tag{3.5}$$

where $f(\dot{\gamma}) \leq 1$ and $\lim_{\dot{\gamma} \to \infty} f(\dot{\gamma}) = 1$.

In other words, the function $\eta_{eff}(T, \dot{\gamma})$ is "split" into two cofactors, the first being dependent only on T, the second on $\dot{\gamma}$. On the basis of the experimental data available, it follows that:

$$\eta_\infty(T) = \eta_\infty(T_0) \exp\left[\frac{E}{R}\left(\frac{1}{T} - \frac{1}{T_0}\right)\right] \tag{3.6}$$

where E is the activation energy for viscous flow and $R = 2 \times 10^{-3}$ kcal/mole is the gas constant.

Equation (13) implies that the Arrhenius law is satisfied (e.g. [56]) which is in good agreement with the published data on plastisols [49–51, 54].

For plastisol I: $E = 8.4$ kcal/mole at $T_0 = 21$ °C, $\eta_\infty(T_0) \simeq 3.5$ Pa · s; for plastisol II: $E = 5.3$ kcal/mole at $T_0 = 20$ °C, $\eta_\infty(T_0) = 5$ Pa · s, $f(\dot{\gamma}) = \eta_{eff}(T_0, \dot{\gamma})/5$.

3.3 Time Dependence of Plastisol Viscosity in the Course of Gelatination at Different Temperatures

The data obtained from studies [34-36, 54, 55] carried out at fixed $\dot{\gamma}$ and various T (as in Figs. 5 and 6[1] show that the time dependence of η_{eff} may be approximated by a linear law. The influence of medium deformation on gelatination can not be determined within the limits of experimental uncertainty. This may be seen, for instance, from Fig. 6a where the dependence $\eta_{eff}(\tau)$ at 53 °C was obtained for both the discrete (points 1) and continuous deforming modes (points 2). Figure 6 presents only those η_{eff} which correspond to the moments of discrete measurements.

The experiments show that the time-temperature dependence of viscosity in the non-Newtonian region of flow may be represented as:

$$\eta_{eff} = \eta_\infty(\tau, T) f(\dot{\gamma}), \qquad \lim_{\dot{\gamma} \to \infty} \eta_{eff} = \eta_\infty(\tau, T) \qquad (3.7)$$

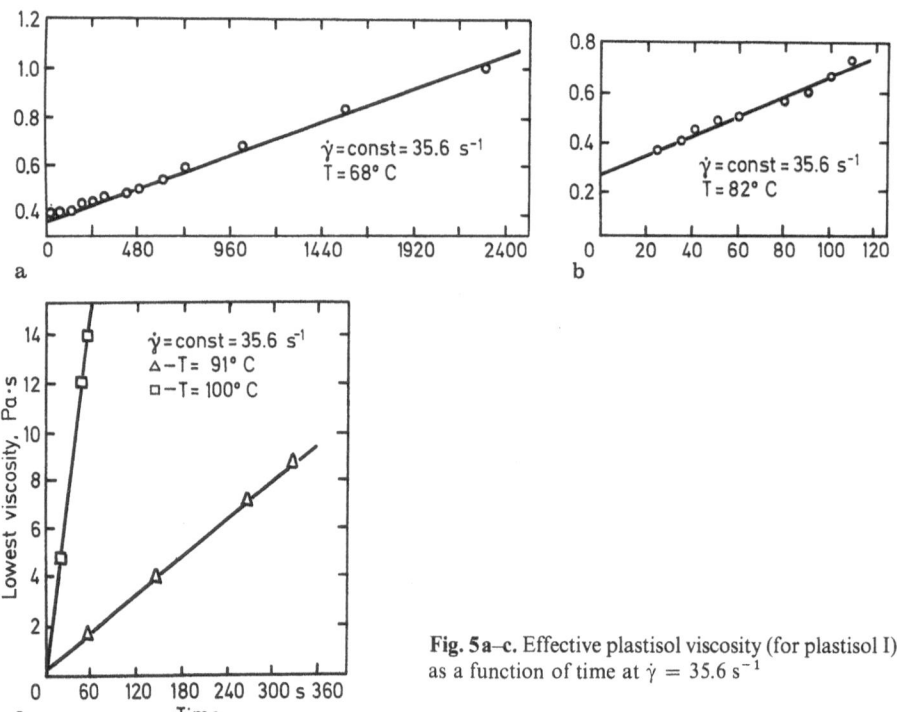

Fig. 5a–c. Effective plastisol viscosity (for plastisol I) as a function of time at $\dot{\gamma} = 35.6 \text{ s}^{-1}$

---

[1] Shear rates $\dot{\gamma}$ for plastisol I were arbitrarily chosen within the region of Newtonian flow; for plastisol II — from consider that the deformation would be rather strong but that the effect of edge "flaking" of the material being structured in the clearance between the cone and plane of the rheogoniometer would not occur.

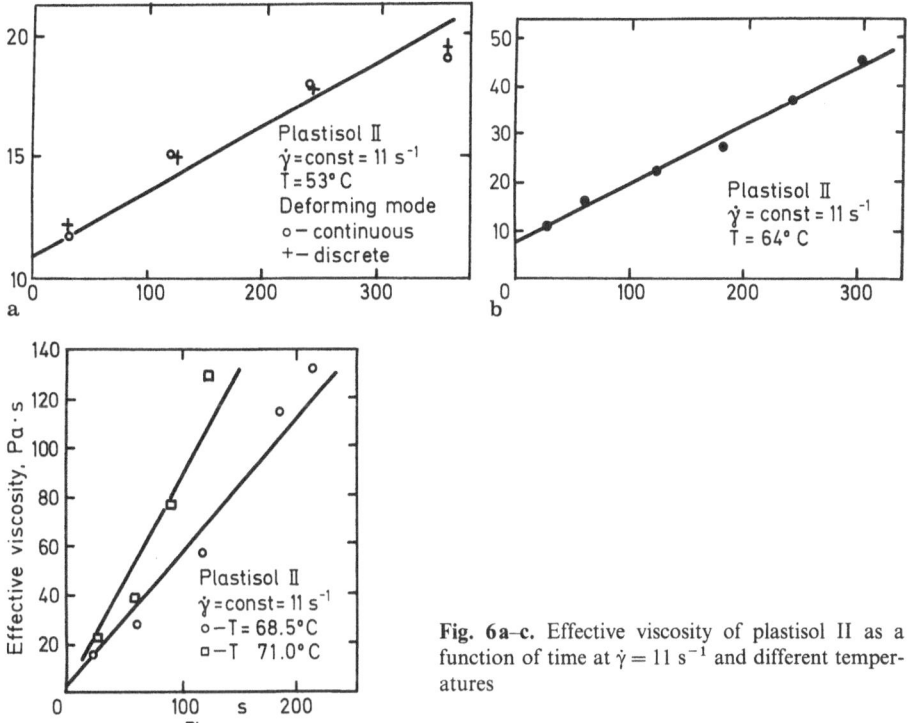

c

Fig. 6a–c. Effective viscosity of plastisol II as a function of time at $\dot\gamma = 11\ \text{s}^{-1}$ and different temperatures

In the temperature range of negligible gelatination, Eq. (3.7) transforms into Eq. (3.5), the function $f(\dot\gamma)$ being the same in Eqs. (3.5) and (3.7). At T $\pm$ 53 °C (for plastisol II) when gelatination is essential, Eq. (3.7) follows from the data of Fig. 7, since $\eta_{\text{eff}}(\tau)$ for $\dot\gamma = 11\ \text{s}^{-1}$ and $1.1\ \text{s}^{-1}$ are given by two parallel straight lines (Fig. 7, points 1 and 2). As follows from the experimental data [36, 54, 55] and Eq. (3.7), the dependence of the lowest viscosity η_∞ (at fixed temperatures) on time with the

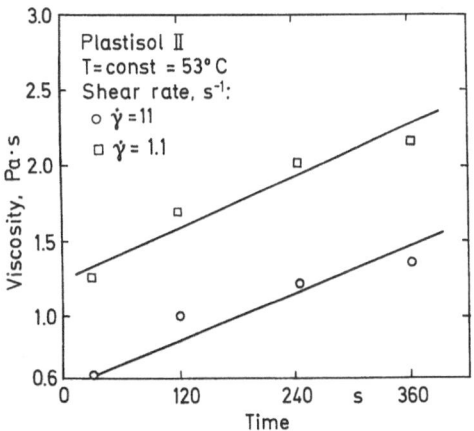

Fig. 7. Time dependence of plastisol II viscosity in the course of gelatination and various shear rates

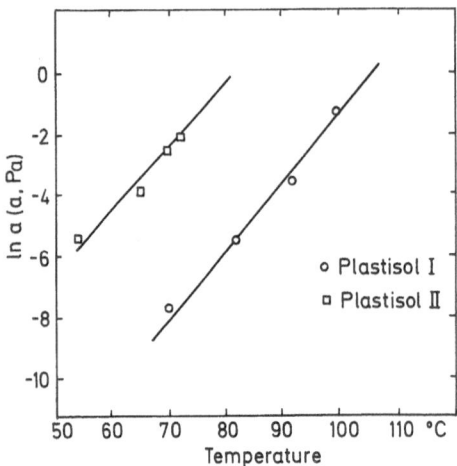

Fig. 8. Temperature dependence of the parameter in Eq. (3.8) in semilog coordinates for plastisols I and II

accuracy sufficient for the purposes of engineering of plastisols may be represented as follows:

$$\eta_\infty = a(T)\,\tau + b(T) \tag{3.8}$$

where T is temperature in K, $a(T) = $ const and $b(T) = $ const at the given temperature.

For real processes, it may be assumed that at $T \lesssim 40\text{--}60\ ^\circ C$ $a(T) \simeq 0$ (gelatination practically does not occur) and at $T \gtrsim 40\text{--}60\ ^\circ C$ $a(T) \neq 0$.

The function $\ln a(T)$ for both plastisols are given in Fig. 8. They represent straight lines, from which it follows that:

$$a(T) = a(T_0)\,\exp\,[m(T - T_0)] \tag{3.9}$$

where T_0 is some fixed temperature, m and $a(T_0)$ are constants. For both plastisols, the dimensionless constant $m = 0.22$, but the values of $a(T)$ are different. For instance, $a(T_0) = 3.4 \times 10^{-4}$ Pa at $T_0 = 68\ ^\circ C$ (for plastisol I) and $a(T_0) = 0.1$ Pa (for plastisol II).

Now let's determine the dependence $b(T)$. At $60\ ^\circ C$ for paste I and $T \lesssim 45\ ^\circ C$ for paste II, the value of b is equal to the lowest Newtonian viscosity at any moment of time. At temperatures above the stated values, the value of b was determined from the intercepts of straight lines $\eta_\infty(\tau)$ with the ordinate axes of the plots of the types shown in Figs. 5–7. The experimentally determined values $b(1/T)$ at any temperature fit the same linear plot (Fig. 9). Hence, the dependence $b(T)$ is given by Eq. (3.6), provided that $\eta_\infty \leftrightarrow b$ is substituted by b, constants remaining the same as in the case of stationary flow.

At $(T - T_0)/T_0 \ll 1$ eq, Eq. (13) may be rewritten as:

$$b(T) = b(T_0)\,\exp\,(-K(T - T_0)]\,, \quad K = E/RT_0^2 \tag{3.10}$$

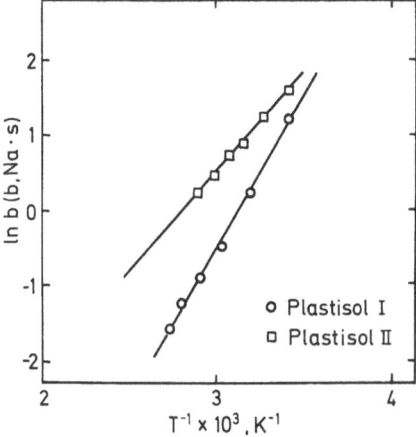

○ Plastisol I

□ Plastisol II

Fig. 9. Temperature dependence of the parameter $a(T)$ in Eq. (3.8) in semilog coordinates for plastisols I and II

At $T_0 = 68\,°C$, $K = 3.61 \times 10^{-2}$, $b(T_0) = 0.49\,Pa \cdot s$ for material I; and $K = 2.28 \times 10^{-2}$, $b(T_0) = 1.4\,Pa \cdot s$ for material II.

From Eqs. (3.9) and (3.10) it follows that the $b(T)$ decrease with T is slower than the corresponding increase of $a(T)$. For instance, the temperature increase from 92 to 100 °C for plastisol I leads to a 1.3-fold drop of $b(T)$, while $a(T)$ increases by as much as a factor of 9.

Now let's summarize the data on the viscosity of PVC plastisols. First, as concerning injection moulding (filling of thin moulds), both the plastisols (those showing and those not showing thyxotropic phenomena) may be considered as non-linear liquids with no characteristic times provided that the filling time is much larger than the characteristic time of the medium.

Second, the properties of such liquids are well determined by the stationary dependence of effective viscosity η_{eff} on shear rate $\dot{\gamma}$ (at large $\dot{\gamma}$, the lowest Newtonian viscosity is attained at $T \lesssim 40\text{--}60\,°C$). At higher temperatures, together with $\eta_{eff}(\dot{\gamma})$, one has to known the dependence of η_{∞} on time as well. This dependence is linear, and its coefficients are exponentially dependent on temperature.

Third, the rheological properties of plastisols during isothermal filling of thin moulds (the applicability of isothermal injection into thin flat moulds will be referred to below) at some medium range $\dot{\gamma}$ (corresponding to the flow with η_{∞}) and considerations of dimensions suggest that the filling time for a flat mould, t_m, as a function of pressure may be represented as:

$$t_m = \frac{b}{P} f\left(\frac{a}{P}\right), \qquad \lim_{x \to 0} f(x) = c \qquad (3.11)$$

where $f(x)$ is some monotonically increasing function and C is the constant determined by the geometry of a mould.

If gelatination is negligible (e.g. filling cold moulds), Eq. (3.11) will be reduced to:

$$t_m = \frac{b}{P}. \qquad (3.12)$$

4 Theory of Mould Filling with Plastisols

Figure 10 shows the simplest filling geometries to be analysed in this section:
— flat mould with a slit runner (Fig. 10a);
— flat cavity with a "point" runner (Fig. 10b);
— flat mould with a "point" runner (Fig. 10c).

Though these schemes may serve as simplified models for more complicated forming equipment, in practice one may encounter very diverse configurations (corresponding to those of finished articles). Calculations for these complicated moulds are, of course, more sophisticated and require computers. Complication arise due to variable cross sections of forming cavities when $\delta = \delta(z)$ and $B = B(z)$. Since this section aims at the study on the main features of low-pressure injection moulding, we shall analyse the three geometries in detail outlining the physical meaning of the phenomena occurring at each stage of injection moulding.

In the further discussion, the effects of inertia will be neglected. This simplifying assumption is based on the following estimations. As an example of making thin and long articles, let's assume that the characteristic linear velocity of filling $U_f \simeq 10$ cm/s, the length of cavity $H \simeq 100$ cm, the slit width $\delta \simeq 0.1$ cm. Then the characteristic time of filling $t_f \simeq H/U_f \simeq 10$ s, and the characteristic shear rate during filling $\dot{\gamma} \simeq U_f/\delta \simeq 100$ s^{-1}. As was shown in Sect. 3, at such $\dot{\gamma}$ plastisols are practically in the rheological state which is characterized by the viscous Newtonian flow with the second (lowest) Newtonian viscosity $\eta = \eta_\infty$. In Sect. 3.2, the value of η_∞ was

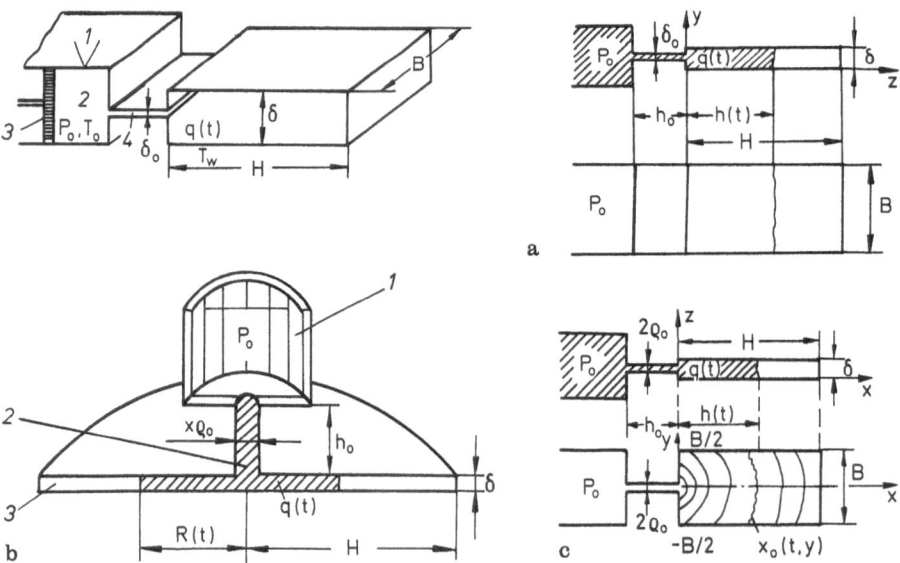

Fig. 10a–c. Existing (I) and proposed (II) principal schematic diagrams of plastisol moulding: **a** into a flat cavity through a slit runner: 1 — feeding funnel; 2 — receiver; 3 — plunger; 4 — slit runner; 5 — forming clearance; **b** into a fan mould through a point runner; 1 — plastisol receiver; 2 — runner; 3 — radially-flat cavity; **c** into flat slit mould through a point runner

shown to be equal to ~ 50 P (for platisols of medium viscosity); in such a case the Reynolds number for cavity filling will be:

$$Re = \frac{dU_f \, \delta}{\eta} \approx \frac{1 \times 10 \times 10^{-1}}{50} \sim 2 \times 10^{-2} \ll 1$$

where $d \simeq 1$ g/cm^3 is plastisol density.

The obtained Reynolds number R_e gives an upper estimation for real plastisols and shows that in designing filling processes all the effects of inertia may be neglected with no essential losses in accuracy. This is mainly due only to the small thickness δ of forming cavities (in making the most of articles, films, coatings, clothing accessories, etc.).

First of all, we shall consider the filling with Newtonian liquids (which is typical for a number of real plastisols of type I, cf. Sect. 3.2 and 3.3) and analyze the case of isothermic filling. Moreover, gelatination will still be neglected which holds true for low-temperature moulds of for very fast filling. Then we shall proceed to more complicated cases, with the account of some additional effects, which will provide the completeness and workability of theory.

4.1 Filling a Flat Mold through a Slit Gate (Runner)

The characteristic dimensions shown in Fig. 10a satisfy the following inequalities:

$$\delta_0 < \delta \ll H, \qquad B < H, \qquad \delta \ll B[(\delta_0/h_0) \ll 1, \, h_0 \ll H)] \qquad (4.1)$$

The fulfillment of the two latter inequalities in parentheses is not necessary in practice. In Fig. 10a, h(t) denotes the flow front which at the first approximation will be assumed to be flat, as elaborated in [56]. This is known [57] to be valid under the conditions $Re \ll 1$ and at the absence of the "jet" effect [57-60].

In the simplest case under consideration (Newtonian, isothermal, and inertialess motion in the absence of structuring) at $h(t) \gg \delta$ the flow may be assumed to be quasi-one-dimensional, and the distorsions of velocity and pressure profiles in the vicinity of a front and the gate into a forming cavity may be neglected. The basic equation will then have the form:

$$\frac{1}{\eta} \frac{\partial P}{\partial z} = \frac{\partial^2 v}{\partial y^2}; \qquad v|_{y=0} = v|_{y=\delta} = 0 \qquad (4.2)$$

where $v = v(t, y)$ is the velocity of plastisol motion in the direction of filling (along the z axis), and $P(z, t)$ is the current pressure in a forming cavity.

The velocity profile has the Poiseuille form:

$$v = -\frac{y(\delta - y)}{2\eta} \frac{\partial P}{\partial z},$$

and the rate of flow (per unit of slit width) is given by:

$$Q = Q(t) = \int_0^\delta v \, dy = -\frac{\delta^3}{12} \frac{\partial P}{\partial z} \qquad (4.3)$$

The region occupied by moving plastisol is determined by the inequality $0 < z < h(t)$, $0 < y < \delta$; at the flow front, $z = h(t)$ and thus:

$$\langle v \rangle = \frac{Q}{\delta} = -\frac{\delta^2}{12\eta}\frac{\partial P}{\partial z}\bigg|_{z=h(t)} = \frac{dh}{dt}, \qquad P|_{z=h(t)} = 0 \tag{4.4}$$

The first of these conditions is kinematic, the second is dynamic in nature. $\langle\ \rangle$ denotes averaging over the slit width.

If the pressure in the slit cavity ($z = 0$) is denoted by $q(t)$, then from Eq. (4.3) and the second condition in Eq. (4.4) we shall find that at each moment the pressure distribution for the Newtonian plastisol is linear:

$$P(z, t) = q(t)\left[1 - \frac{z}{h(t)}\right], \qquad -\frac{\partial P}{\partial z} = \frac{q(t)}{h(t)} \tag{4.5}$$

From Eqs. (4.3)–(4.5) it follows that:

$$\frac{dh}{dt} = \frac{Q}{\delta}\frac{\delta_0^2}{12\eta}\frac{q(t)}{h(t)} \tag{4.6}$$

For $q(t)$ determination, now consider the equation of flow rate (mass balance) through the entrance (runner) channel corresponding to the Poiseuille formula for a flat channel:

$$Q(t) = \frac{\delta^3}{12\eta}\frac{P_0 - q(t)}{h_0} \tag{4.7}$$

where $P_0 = P_0(t)$ is the pressure in a cylinder ("injection pressure") which generally is time dependent, but in practice the regime $P_0(t) = \text{const}$ is normally used.

Comparing Eq. (4.6) with Eq. (4.7) we find:

$$q(t) = \frac{P_0 h(t)}{h(t) + h_0(\delta/\delta_0)^3}; \tag{4.7a}$$

Substituting Eq. (4.7a) into Eq. (4.6), we obtain:

$$\frac{dh}{dt} = \frac{\delta^2 P_0/12\eta}{h(t) + h_0(\delta/\delta_0)^3}; \qquad h(0) = 0 \tag{4.8}$$

The initial condition in Eq. (4.8) has an asymptotic character, since, as noted previously, Eq. (4.8) is valid only if $h(t) \gg \delta$. Equation (4.8) may be readily solved for any function $P_0 = P_0(t)$. For the practically important case of $P_0 = \text{const}$, the solution to Eq. (4.8) has the form:

$$\frac{h(t)}{h_0}\cdot\left(\frac{\delta_0}{\delta}\right)^3 = \sqrt{1 + \left(\frac{\delta}{h_0}\right)^2\left(\frac{\delta_0}{\delta}\right)^6\frac{P_0 t}{6\eta}} - 1 \tag{4.9}$$

Equation (4.9) provides a possibility to determine all important parameters (from the technological point of view) of mould filling:
— filling time

$$t_f = t|_{h=H} = \frac{6\eta H}{\delta^2 P_0}[H + 2h_0(\delta/\delta_0)^3]$$ (4.10)

— filling rate

$$1 = \frac{\delta_0^3}{12\eta h_0} \cdot \frac{P_0/\delta_2}{\sqrt{1 + \left(\frac{\delta}{h_0}\right)^2 \left(\frac{\delta_0}{\delta}\right)^6 \frac{P_0 t}{6\eta}}} \cdot$$

— mean filling rate

$$1 = \frac{1}{t_f}\int\limits_0^{t_f} 1\,dt_f = \frac{H}{t_f} = \frac{\delta^2 P_0/6\eta}{H + 2h_0(\delta/\delta_0)^3}$$ (4.11)

Equations (4.5) and (4.7a) imply that the pressure distribution has the form:

$$P(z, t) = \frac{P_0[h(t) - z]}{h(t) + h_0(\delta/\delta_0)^3} \quad (0 < z < h(t))$$ (4.12)

from which it follows that, for the fixed point $z < h(t)$ within a cavity, pressure grows constantly with time.

Equation (4.10) deserves the special attention of technologists since it has one important particular feature: there exist some $\delta = \delta_*$ for which the filling time t_f is *minimal* and

$$\delta_* = \delta_0(H/h_0)^{1/3}$$ (4.13)

The existence of such δ_* has a clear physical meaning. As is seen from Eq. (4.5), the driving force of filling is the pressure $q(t)$ at the entrance into a cavity. If, for instance, $q(t) = $ const, then upon δ increase the resistance to the flow in a cavity decreases, and t_f decreases as well. However, the value of $q(t)$ per se also monotonically increases from $q(0) = 0$, this increase being faster, the lower the value of δ and the larger the resistance to flow; liquid is likely to form a "plug" at the entrance to the cavity. This double role of the thickness of a forming cavity is the reason for the existence of minimal time $(t_f)_{min}$. Of course, the role of the runner channel is also very important. If one out of the four geometrical parameters δ, δ_0, H, and h_0 is arbitrarily chosen, then it will have to be taken according to Eq. (4.13); then we have:

$$(t_f)_{min} = \frac{18\eta}{P_0} \frac{H^{4/3}h_0^{2/3}}{\delta_0^2}$$ (4.10a)

As is seen, t_f increases together with η, H, and h_0, and decreases with P_0 and δ_0, which is clear from both the physical and technological points of view.

4.2 Filling a Fan Mould Through a "Point" Gate (Runner)

In this case $R(t)$ is the coordinate of the flow front (for the sake of simplicity represented as concentric circles), and the characteristic dimensions of a cavity (Fig. 10b) satisfy the inequality $\varrho_0 < \delta \ll H$ ($\varrho_0 \ll h_0$), the condition in parentheses, just as in Eq. (4.1), is not necessary.

Now let's adopt the hypothesis on purely radial flow which will hold true if $R(t) \gg \varrho_0$ (see Fig. 10b). The equation of flow will have the form:

$$
\left.
\begin{array}{l}
\dfrac{1}{\eta} \cdot \dfrac{\partial P}{\partial \varrho} = \left(\dfrac{\partial^2}{\partial \varrho^2} + \dfrac{1}{\varrho} \cdot \dfrac{\partial}{\partial \varrho} - \dfrac{1}{\varrho^2} + \dfrac{\partial^2}{\partial y^2} \right) v ; \qquad \dfrac{\partial P}{\partial y} = 0 ; \qquad \dfrac{\partial}{\partial P} (\varrho v) = 0 \\[4mm]
P_{\varrho\varrho} = -P + 2\eta \dfrac{\partial v}{\partial \varrho} ; \qquad P_{\varphi\varphi} = -P + 2\eta \dfrac{v}{\varrho} ; \qquad P_{\varrho y} = \eta \dfrac{\partial v}{\partial y}
\end{array}
\right\}
$$

$$(4.14)$$

where P is pressure and $v \equiv v_\varrho$ is the radial velocity of the component of the stress tensor.

At the entrance into a fan cavity, the next boundary condition:

$$P|_{\varrho=\varrho_0} = q(t) \tag{4.15}$$

and kinematic and dynamic conditions at the filling front:

$$\langle v \rangle|_{\varrho=R(t)} = \frac{dR}{dt} ; \qquad \langle P_{\varrho\varrho} \rangle|_{\varrho=R(t)} = 0 \tag{4.16}$$

should be satisfied. Just as before, $\langle \ \rangle$ denotes averaging over the slit width. The condition of adherence to the slit walls should also be fulfilled:

$$v|_{y=0} = v|_{y=\delta} = 0 \tag{4.17}$$

The solution of the problem found in Eqs. (4.14)–(4.17) is described elsewhere [34, 36, 54], here we shall present only the most important intermediate and final expressions.

The desired pressure $q(t)$ at the entrance into a forming cavity (determined by analogy with Sect. 4.1) is given by:

$$q(t) = \frac{P_0(t) \ln [R(t)/\varrho_0]}{\dfrac{4}{3} \dfrac{\delta^3 h_0}{\varrho_0^4} + \ln [R(t)/\varrho_0]} \tag{4.18}$$

At $R(t) \gg \varrho_0$ for the case $P_0(t) = \mathrm{const}$, we obtain the solution:

$$\left(\frac{8}{3} \frac{\delta^3 h_0}{\varrho_0^4} - 1 \right) \left[\left(\frac{R(t)}{\varrho_0} \right)^2 - 1 \right] + 2 \left(\frac{R(t)}{\varrho_0} \right)^2 \ln \left(\frac{R(t)}{\varrho_0} \right) = \frac{P_0 t}{3\eta} \left(\frac{\delta}{\varrho_0} \right)^2 \tag{4.19}$$

Assuming $R(t_f) = H$ in Eq. (4.19) and taking t_f for the filling time, we obtain:

$$t_f \approx \frac{6\eta}{P_0} \left(\frac{H^2}{\delta}\right) \left[\ln\left(\frac{H}{\varrho_0}\right) + \frac{4}{3}\frac{\delta^3 h}{\varrho_4} - \frac{1}{2}\right]. \qquad (4.20)$$

The mean filling velocity has the form:

$$\bar{R} = \frac{1}{t_f} \int\limits_0^{t_f} \dot{R}\, dt = \frac{H - \varrho_0}{t_f} \approx \frac{H}{t_f} \qquad (4.21)$$

It should be outlined that, just as in the flat case, there exist such δ_* at which $t_f = (t_f)_{min}$:

$$\delta_* = \varrho_0 \left[\frac{3}{2}\frac{\varrho_0}{h_0} \ln\left(\frac{H}{\varrho_0}\right)\right]^{1/3} \qquad (4.22)$$

If this correlation is acceptable, then:

$$(t_f)_{min} = \frac{18\eta}{P_0} \frac{H/\varrho_0[\ln(H/\varrho_0) - 1/6]}{[3/2P_0/h_0 \ln(H/\varrho_0)]^{2/3}} \qquad (4.23)$$

Equation (4.23) holds true when $H/\varrho_0 \gg e^{1/6}$. The existance of δ_*, which minimizes the filling time for radial forming cavity, is conditioned by the same physical (hydrodynamical) reasons as in the case of a flat moulding.

4.3 Filling a two-dimensional Flat Mould Through a Point Gate (Runner)

For this more complicated case (filling scheme shown in Fig. 10c), a filling cavity will be assumed to possess an axis of symmetry, i.e. $-B/2 < y < B/2$, and the reasonable inequalities are satisfied:

$$\varrho_0 < \delta \ll H; \qquad \delta \ll B, \qquad B < H(\varrho_0/l_0 \ll 1, l_0 \ll H) \qquad (4.24)$$

(The latter inequalities in parentheses are not necessary in practice).

Just as before, we shall consider the inertialess, Newtonian isothermal flow and make two additional assumptions simplifying the solution:

(1) liquid flow along the z axis is neglected, i.e. $v_z = 0$ (it is acceptable since we consider a two-dimensional *flat* mould);

(2) change in longtitudinal velocities v_x and v_y (along the x and y axes) is much more slow than along the transverse z coordinate.

Let's note that assumptions (1) and (2) are fulfilled only at some time interval after mould filling.

In the region from a "point" gate (entrance) to a flow front, the equation of motion may be written as:

$$\frac{1}{\eta}\frac{\partial P}{\partial x} = \frac{\partial^2 v_x}{\partial z^2}; \qquad \frac{1}{\eta}\frac{\partial P}{\partial y} = \frac{\partial^2 v_y}{\partial x^2}; \qquad \frac{\partial v}{\partial x} + \frac{\partial v}{\partial y} = 0 \qquad (4.25)$$

With the conditions of adherance:

$$v_x|_{z=0} = v_y|_{z=0} = v_x|_{x=\delta} = v_y|_{x=0} = 0$$

taken into account, we obtain the following solutions:

$$v_x = -\frac{z(\delta - z)}{2\eta}\frac{\partial P}{\partial x}; \qquad v_y = -\frac{z(\delta - z)}{2\eta}\frac{\partial P}{\partial y}$$

The respective "flows" along the x and y directions are as follows:

$$\left.\begin{array}{l} Q_x \equiv U = \displaystyle\int_0^\delta v_x\,dz = -\frac{\delta^3}{12\eta}\frac{\partial P}{\partial x} \\[2em] Q_y \equiv V = \displaystyle\int_0^\delta v_y\,dz = -\frac{\delta^3}{12\eta}\frac{\partial P}{\partial y} \end{array}\right\} \tag{4.26}$$

From the continuity conditions in Eq. (4.25), it follows that:

$$\frac{\partial U}{\partial x} + \frac{\partial V}{\partial y} = 0,$$

which indicates that the pressure P satisfies the two-dimensional Laplace equation:

$$\Delta P \equiv \frac{\partial^2 P}{\partial x^2} + \frac{\partial^2 P}{\partial y^2} = 0 \tag{4.27}$$

which infers that P is a harmonic function. Equations (4.26 and 4.27) reduce the problem of cavity filling to the flat problem with unknown boundary. In this problem, all the boundary conditions are written at several plane contours: $\overline{1}$ describing the side walls $y = \pm B/2$ and the portions of the front side $x = 0$ ($|y| > \varrho_0/2$); $\overline{2}$ describing entrance into a cavity ($x = 0$, $|y| < \varrho_0/2$), and $\overline{0}$ describing the flow front $x_0 = x_0(y, t)$.

As for contour $\overline{1}$, in view of assumption (2) in the first two equations of Eq. (4.25) at motionless boundaries we may impose only the conditions of "impenetrability" which means cancelling the flow normal to the wall, i.e.:

$$Q_{\vec{n}}|_{\overline{1}=0}(Q_y \equiv V|_{y = \pm B/2}) = 0; \qquad Q_x \equiv U|_{x = 0, |x| < \varrho_0} > 0 \tag{4.28}$$

where $\vec{n}$ is a unit vector normal to the moving boundary $x_0(y, t)$.

The dynamic condition at the entrance into a cavity (at contour $\overline{2}$) has the form:

$$P_{\overline{2}} = q(t) \qquad (\overline{2}: \varrho = \sqrt{x^2 + y^2}) \tag{4.29}$$

where $\overline{2}$ is an arc of the circle of radius ϱ_0.

The dynamic condition at a moving (unknown) boundary given by $x = x_0(y, t)$ has the form:

$$P|_{x = x_0(y, t)} = 0 \tag{4.30}$$

From the analysis of the kinematic conditions at a moving boundary [36], with Eq. (4.26) taken into account, we obtain:

$$\frac{\partial x_0}{\partial t} = \frac{\delta^2}{12\eta} \frac{\partial x}{\partial P}\bigg|_{x_0} ; \qquad \frac{\partial x_0}{\partial t} \cdot \frac{\partial x_0}{\partial y} = -\frac{\delta^2}{12\eta} \frac{\partial P}{\partial y}\bigg|_{x0} \tag{4.31}$$

Therefore, Eq. (4.27) together with the boundary conditions of Eqs. (4.28–4.31) provide a definition of the problem of pressure $P(x, y, t)$ with an unknown boundary $x_0(y, t)$ provided that function $q(t)$ specifying the pressure at the exit from a "point" gat into a cavity is known. In practice, function $P_0(t)$ is known; having determined the mass balance function $q(t)$, the final formulation takes the form:

$$\left.\begin{array}{l}
\Delta P = 0; \qquad P|_{\sqrt{x^2 + y^2} = \varrho_0} = P_0(t) - \dfrac{16 \cdot h_0 \delta \eta}{\pi \varrho_0^4} \displaystyle\int_0^{B/2} x_0(y, t)\, dy \\[18pt]
\dfrac{\partial P}{\partial n}\bigg|_{\Gamma_1} = 0; \qquad P|_{x_0} = 0; \qquad \dfrac{\partial P}{\partial x}\bigg|_{x_0} + \dfrac{12\eta}{\delta^2} \dfrac{\partial x_0}{\partial t} = 0 \\[18pt]
\dfrac{\partial P}{\partial y}\bigg|_{x_0} \dfrac{12\eta}{\delta} \dfrac{\partial x_0}{\partial y} \dfrac{\partial x_0}{\partial t} = 0
\end{array}\right\} \tag{4.32}$$

Since the influence of walls on the U and V flow components disappears upon the separation of the order of δ from a solid wall, inaccuracy of the scheme cannot have an essential influence on the flow rate calculations. Equations (4.32) may have an explicit solution within some limited time interval when the flow from a "point" runner is still radial, while the flow front is circular. This filling pattern is realized up to the moment when the radius of a free boundary $R(t)$ will become equal to $B/2$ (see Fig. 11). Such a filling regime we shall refer to as the *first stage of filling*; the formulae of 4.2 with some minor corrections (for details see [36]) are applicable to this stage; the corrections arise mainly due to the fact that the radial flow occurs in

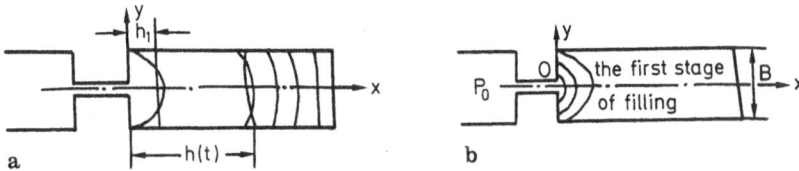

a   b

Fig. 11a, b. Kinetic identification of the first stage of filling a flat mould through a round (point) runner: **a** flow is purely radial during some limited time interval, flow front is approximately circular; **b** the radius of free boundary becomes equal to half-width of the forming cavity $R(t) = B/2$

the sector with the φ angular coordinate varying within the range $(0, \pi)$, whereas in the preceding section the radial flow occurred within the entire interval φ $(0, 2\pi)$.

At $P_0 = $ const, the solution to Eqs. (4.32) has the form [compare with Eq. (4.19)]:

$$\left(\frac{4}{3}\frac{\delta^3 h_0}{\varrho_0^4} - 1\right)\left[\left(\frac{R(t)}{\varrho_0}\right)^2 - 1\right] + 2\left(\frac{R(t)}{\varrho_0}\right)^2 \ln\left(\frac{R(t)}{\varrho_0}\right) = \frac{P_0 t}{3\eta}\left(\frac{\delta}{\varrho_0}\right)^2 \quad (4.33)$$

Filling time for the first stage is determined from the condition $R(t_f^1) = B/2$. From Eq. (4.33) with account of $(B/2\varrho_0) \gg 1$, we obtain:

$$t_f^1 \approx \frac{6\eta}{P_0}\left(\frac{B}{2\delta}\right)^2\left[\ln\left(\frac{B}{2\varrho_0}\right) + \frac{2}{3}\frac{\delta^3 h_0}{\varrho_0^4} - 1/2\right] \quad (4.34)$$

The second stage of filling begins when the surface of the filling front undergoes constraining action of the side walls (with coordinates $y = \pm B/2$). The filling front is levelled and at some distance from the entrance into the forming cavity becomes almost planar. The second stage may also be described by the boundary conditions of Eqs. (4.32), the solution of which may be obtained by computation. The approximate solution based on averaging all the equations over the y variable (Fig. 11) has been reported by the authors [36, 54]. The averaged pressure was shown to be the linear function of x, and after some transformations (the vanishing of mean pressure upon averaging over y, the filling front taken into account) the expression describing the second stage of filling was obtained:

$$\frac{P_0\delta^2}{12\eta}(t - t_f^1) = \frac{h(t) - B\pi/8}{2}\left[h(t) + B\left(\frac{\pi}{8} + \frac{4}{3\pi}\frac{h_0\delta^3}{\varrho_0^4}\right)\right] \quad (4.35)$$

The second stage of filling is completed when the central front of flow $h(t)$ reaches the bottom of a forming cavity (H). Then from Eq. (4.35) we have:

$$t_f^2 = \frac{6\eta}{P_0\delta^2}(H - B\pi/8)\left[H + B\left(\frac{\pi}{8} + \frac{4}{3\pi}\frac{h_0\delta^3}{\varrho_0^4}\right)\right] \quad (4.36)$$

Defining the mould filling time as a sum of times of both the stages $t = t_f^1 + t_f^2$ and making use of Eqs. (4.34) and (4.36), we obtain:

$$t_f = \frac{B\eta}{P_0\delta^2}\left[H^2 + \frac{4}{3\pi}BH\frac{h_0\delta^3}{\varrho_0^4} + \frac{B^2}{4}\left(\ln\frac{B}{2\varrho_0} - \frac{\pi^2}{16} - \frac{1}{2}\right)\right] \quad (4.37)$$

Now let us discuss shortly the effect of geometry on t_f. First, in view of the assumption $B/\varrho_0 \gg 1$ the third term in parentheses of the right-hand side of Eq. (4.37) will be positive (which is satisfied if $(B/2\varrho_0) > 3.127$; for "point" runners this normally holds true). Then t_f grows together with increasing H, B, h_0 and decreasing the radius of the runner channel ϱ_0. Second, an analysis of Eq. (4.37) shows that there exist such $\delta = \delta_*$ for which t_f is minimal. The expression for δ_* is as follows:

$$\delta_* = \left\{\frac{3\pi\varrho_0^4}{4h_0}\left[2H^2 + \frac{B^2}{2}\left(\ln\frac{B}{2\varrho_0} - \frac{\pi^2}{16} - \frac{1}{2}\right)\right]\right\}^{1/3} \quad (4.38)$$

and correspondingly:

$$(t_f)_{min} = \frac{18\eta}{P_0\delta_*^2}\left[H^2 + \frac{B^2}{4}\left(\ln\frac{B}{2\varrho_0} - \frac{\pi^2}{16} - \frac{1}{2}\right)\right] \tag{4.39}$$

Therefore, if it turns out to be possible to vary the dimensions of a forming cavity until Eq. (4.38) be fulfilled, it will have to be done, since it leads to minimization of filling time (and the process as a whole) and hence to enchanced productivity of the injection moulding of plastisols.

The following sections are devoted to the analysis of some specific phenomena which we neglected in the previous sections.

4.4 The Effect of Pseudo-plasticity in Filling Injection Moulds with Plastisols

The effect of pseudo-plasticity was reportedly known for different plastisols, in particular, for plastisols II described in Sect. 3. The rheological behaviour of these systems with an accuracy sufficient for the purposes of engineering is given by the Bingham-Shvedov law within the extended range of shear rates ($10^{-2} < \dot\gamma < 10^2$):

$$\begin{matrix} \tau = \tau_0 + \eta_{pl}\cdot\dot\gamma & (|\tau| > \tau_0) \\ \dot\gamma = 0 & (|\tau| < \tau_0) \end{matrix} \Big\} \tag{4.40}$$

For instance, for the parameters of plastisol II specified in Sect. 3.2 the ultimate shear stress (the so-called "yield point") τ_0 and the plastic viscosity η_{pl} have the values: $\tau_0 = 100$ Pa; $\eta_{pl} = 4$ Pa $\cdot$ s.

Here the effective viscosity is defined as:

$$\eta_{eff} = \frac{\tau_0}{\dot\gamma} + \eta_{pl} \tag{4.40a}$$

as is shown by points in Fig. 5a.

As an example of utilizing Eq. (4.40), let's consider the filling of a flat cavity with a slit gate such as that analyzed in Sect. 4.1 for the case of Newtonian flow.

For flow rate Q per unit area with a cross section of flat tubing with thickness δ we may write the well-known expression:

$$Q = \frac{\delta^2\tau_\omega}{6\eta_{pl}}(1 + 2\xi)(1 - \xi)^2\left(\xi = \tau_0/\tau_\omega; \tau_\omega = \frac{\Delta P}{H}\cdot\frac{\delta}{2}\right) \tag{4.41}$$

where τ_ω is shear stress at the tubing wall ΔP the pressure differential, H the length of tubing. It is clear that at $\xi \to 0$ ($\tau_0 \to 0$) Eq. (4.41) converts into Eq. (4.3).

Since the pressure at the flow front $z = h(t)$ vanishes [cf. hydrodynamic condition in Eq. (4.4)], then in calculating the pressure in a cavity ($0 < z < h(t)$) from Eq. (4.41) one has to put:

$$\tau_\omega^{(hole)} = \frac{q(t)}{h(t)}\cdot\frac{\delta}{2}; \quad \xi^{(hole)} = \tau_0/\tau_\omega^{(hole)} = 2\tau_0 h(t)/[q(t)\,\delta] \tag{4.42}$$

where q(t) is the pressure in a slit cavity in the vicinity of a runner; $z = h(t)$ is the coordinate of the flow front.

Now let's use the averaged kinematic condition at the flow front:

$$\frac{dh}{dt} = \frac{Q(t)}{\delta} \tag{4.43}$$

where $Q(t)$ is given by Eqs. (4.41) and (4.42).

To complete this set of equations, let's use Eq. (4.41) for the flow rate through a flat runner channel where one has to put:

$$\tau_\omega^{(run)} = \frac{P_0(t) - q(t)}{h_0} \frac{\delta_0}{2} ; \qquad \xi^{(run)} = \frac{\tau_0}{\tau_\omega^{(run)}} = \frac{2\tau_0 h_0 \cdot}{[P_0(t) - q(t)]\delta_0} \tag{4.44}$$

The obtained set of equations is as follows:

$$\left.\begin{aligned}
Q(t) &= \frac{\delta^3 q(t)}{12\eta_{pl} h(t)} \left(1 + \frac{4\tau_0 h(t)}{q(t)\,\delta}\right)\left(1 - \frac{2\tau_0 h(t)}{q(t)\,\delta}\right)^2 ; \qquad \frac{dh}{dt} = \frac{Q(t)}{\delta} \\
Q(t) &= \frac{\delta_0^3(P_0(t) - q(t))}{12\eta_{pl} h_0}\left\{1 + \frac{4\tau_0 h_0}{[P_0(t) - q(t)]\delta_0}\right\}\left[1 - \frac{2\tau_0 h_0/\delta_0}{P_0(t) - q(t)}\right]^2
\end{aligned}\right\} \tag{4.45}$$

Even for the case $P_0 = $ const, Eqs. (4.45) must be solved by using numerical calculations.

Now consider two asymptotic cases, of small and large pressures of moulding, making use of Eq. (4.41).

(1) The case *of large pressure differentials along the slit channel*. In such a case $\xi = \tau_0/\tau_\omega \ll 1$. Since $(1 + 2\xi)(1 - \xi)^2 = 1 - 3\xi^2 + 2\xi^3$, then at $\xi \ll 1$ we obtain:

$$Q \approx \frac{\delta^2 \tau_\omega}{6\eta_{pl}}(1 - 3\xi^2) \approx \frac{\delta^2 \tau}{6\eta_{pl}}$$

It may therefore be concluded that the filling process is essentially the same as in the case of a Newtonian liquid of viscosity equal to η_{pl}. Similarly, the other formulae of Sects. 4.1 and 4.4 will also hold true for other configurations of forming cavities, provided that η is replaced by η_{pl}.

(2) *The case of small pressure differentials along the slit channel*. In this case parameter ξ in Eq. (4.41) is close to unity, more explicitly $\xi = 1 - \varepsilon$, where $0 < \varepsilon \leqq 1$. Then Eq. (4.41) is reduced to the following asymptotic expression:

$$Q \approx \frac{\delta^2 \tau_0}{2\eta_{pl}}\left(1 - \frac{\tau_0}{\tau_\omega}\right)^2, \tag{4.46}$$

and the set Eqs. (4.45) may be presented in the asymptotic form:

$$\left.\begin{array}{l} Q(t) \approx \dfrac{\delta^2 \tau_0}{2\eta_{pl}} \left(1 - \dfrac{2\tau_0 h(t)}{\delta q(t)}\right)^2 \approx \dfrac{\delta_0^2 \tau_0}{2\eta_{pl}} \left(1 - \dfrac{2\tau_0 h_0/\delta_0}{P_0(t) - q(t)}\right)^2 \\[4mm] \dfrac{Q(t)}{\delta} = \dfrac{dh}{dt} \end{array}\right\} \qquad (4.47)$$

As shown elsewhere [36, 54], the analysis leads to the following asymptotic expression:

$$q(t) \approx P_0 \frac{h(t)}{h_0 + h(t)}, \qquad (4.48)$$

Substitution of Eq. (4.48) into Eq. (4.47) gives:

$$\frac{dh}{dt} \approx \frac{\delta \tau_0}{2\eta_{pl}} \left\{1 - \frac{2\tau_0}{\delta P_0}[h(t) + h_0]\right\}^2; \qquad h(0) = 0 \qquad (4.49)$$

As follows from Eq. (4.49), if:

$$\frac{2h_0 \tau_0}{\delta} \geqq P_0 \qquad (4.49\,a)$$

then the plastisol will not fill a mould, i.e. will not flow out of the runner. If:

$$\frac{2h_0 \tau_0}{\delta} < P_0 \frac{2(H + h_0)}{\delta} \tau_0 \qquad (4.49\,b)$$

then the plastisol will not fill a mould up to its full length H, but up to some length h given by the expression:

$$h_* = \frac{\delta P_0}{2\tau_0} - h_0$$

At last, if:

$$P_0 > 2(H + h_0)\,\tau_0/\delta \qquad (4.50)$$

the material will fill a mould completely. For the case of Eq. (4.50), the solution to Eq. (4.49) has the form:

$$\frac{h(t)}{\left(1 - \dfrac{2\tau_0}{\delta P_0} h_0\right)\left\{1 - \dfrac{2\tau_0}{\delta P_0}[h(t) + h_0]\right\}} = \frac{\delta \tau_0}{2\eta_{pl}} t \qquad (4.51)$$

Equation (4.50) implies that, under the conditions of Eq. (4.50), the mould filling time is given by:

$$t_f = \frac{2\eta_{pl}}{\delta\tau_0} \frac{H}{\left(1 - \frac{2\tau_0}{\delta P_0}\right)\left[1 - \frac{2\tau_0}{\delta P_0}(H + h_0)\right]} \tag{4.52}$$

If only the inequality of Eq. (4.49b) holds true, the filling up to the mentioned length h_* is performed during infinitely long time. The inequality of Eq. (4.50) determines, very importantly from a practical point of view, the lower limit for injection pressure P_0 at which a cavity will be filled completely. To attain *fast filling* of a cavity with a viscous plastisol (e.g. plastisol II), in view of Eq. (4.50) pressure P_0 must satisfy the "stronger" inequality obtained from Eq. (4.50) and $h_0 \ll H$:

$$P_0 \gg 2H\tau_0/\delta \qquad (P_0 \gtrsim 20H\tau_0/\delta) \tag{4.53}$$

Here it was assumed that $a \gg b$ means $a \pm 10b$. It is Eq. (4.53) which allows us to apply the formulae derived in Sects. 4.1–4.3.

4.5 Non-Isothermicity of Filling

The solution to the simplest problem of two-sided heating of an infinitely large layer of thickness δ and initial temperature T_0 at the surfaces to temperature T_ω ($T_\omega \gg T_0$) has the well-known form:

$$\frac{I - I_\omega}{I_\omega - I_0} = \frac{4}{\pi} \sum_{n=1,3,5} \frac{\sin(n\pi/2)}{n} \exp\left(-\frac{\varkappa n^2 \pi^2 t}{\delta^2}\right) \sin\left(\frac{n\pi y}{\delta}\right) \tag{4.54}$$

where $\varkappa$ is thermal diffusivity; for polymers its typical value is of the order of 10^{-2} cm²/s [61].

The first term of the series provides an estimation for the characteristic heating time of a liquid in a thin cavity ($\delta = 0.1$ cm):

$$t_T \sim \frac{\delta^2}{\pi^2 \varkappa} = \frac{10^{-2}}{10 \cdot 10^{-2}} \approx 0.1(s)$$

Hence, plastisol motion within thin cavities (films, plates, coatings, etc.) may be considered as isothermal at temperatures equal to the mould temperature ($T = T_\omega$) already in 0.1 s; and one may consequently assume that $\eta = \eta_\infty(T_\infty)$.

Strictly speaking, this estimation is valid for a motionless layer, not for flowing liquids. The process of slow quasi-stationary non-isothermal flow of the non-Newtonian liquid in a plane tubing of thickness δ (when dissipative heating may be neglected) has been analysed in detail [36, 53, 54] as a preliminary problem. With no account of gelatination, the effect of temperature on flow velocity will be manifested in the viscosity of a liquid, and the velocity will influence convective heat transfer along a tubing.

At the initial and boundary conditions for temperature T:

$$T|_{t=0} = T|_{z=0} = T_0, \qquad T_{y=0} = T_{y=\delta} = T_\omega \qquad (4.55)$$

and by using the averaged parameters (instead of the convective term $v(v, t)\, dT/dz$, it was adopted that $\langle v \rangle\, \partial T/\partial z$, where $\langle v \rangle = Q(t)/\delta$ and $Q(t)$ is liquid consumption at the given pressure differential), the following equation was obtained:

$$\frac{\partial T}{\partial t} + \langle v \rangle \frac{\partial T}{\partial z} = \varkappa \frac{\partial^2 T}{\partial y^2} \; (\langle v \rangle = Q(t)/\delta), \qquad (4.56)$$

which upon transformations [36)] reduces to the standard heat conduction problem:

$$\frac{\partial T}{\partial \tau} = \varkappa \frac{\partial^2 T}{\partial y^2}; \qquad T_{y=0} = T_{y=\delta} = T_\omega; \qquad T|_{\tau=0} = 0 \qquad (4.57)$$

The solution of Eq. (4.57) is thus reduced to Eq. (4.54), provided that t is substituted by τ (see below or [36)]).

Confining to the first term in Eq. (4.54) (in some short time after heating onset), we obtain:

$$T \approx T_\omega - \frac{4}{\pi} (T_\omega - T_0)\, e^{\frac{-\varkappa \pi^2 \tau}{\delta^2}} \sin (\pi y/\delta) \qquad (4.58)$$

The substitution of variables (t by τ) is as follows:

$$\tau = t - z_0^{-1}[z_0(t) - z]; \qquad z_0(t) = \int_0^t \langle v \rangle\, (t_1)\, dt_1 = \frac{1}{\delta} \int_0^t Q(t_1)\, dt_1 \quad (4.59)$$

where z_0^{-1} is the function inverse to $z_0(t)$.

These formulae represent the closed set of equations [where τ is defined as $\tau(x, t)$ according to Eq. (4.59)] determining the pressure profile $P(z, t)$ and providing a possibility to derive $Q(t)$ as a function of $\Delta P(t)$.

In spite of the simplifying assumptions, the set is still rather complicated. Consequently, it seems reasonable to consider the so-called case of *fast thermal tuning* which is valid and important for many cases of practical heating in slits and channels (recall the estimate $t_T \approx 0.1$ s). In such a case, the function $\tau(z, t)$ defined in Eq. (4.59) may be expanded by x in the vicinity of $z = 0$ (since the distance at which the liquid is heated up to the wall temperature T_ω is small); confining ourselves to the first two terms, we obtain:

$$\tau \approx \frac{z\delta}{Q(t)} + Q(z^2) \qquad (4.60)$$

Equation (4.60) corresponds to discarding $\partial T/\partial t$ from the left-hand side of the equation of convective heat transfer, Eq. (4.56), which has an apparent physical significance for fast thermal tuning.

The analysis and appropriate transformations lead to the following expression for the pressure differential under conditions when Eq. (4.60) is applicable:

$$\frac{\Delta P}{H} = \frac{12\eta_\omega}{\delta_3} Q(t) \left[1 + \frac{24\delta Q}{\varkappa\pi^4 H} \left(1 - \frac{8}{\pi^2} \right) \frac{E}{RT_\omega} \left(1 - \frac{T_0}{T_\omega} \right) \right] \qquad (4.61)$$

If the following inequality is satisfied:

$$\frac{\varkappa\pi^2 H}{Q(t)\,\delta} \gg 1 \qquad (4.62)$$

then the expression in square brackets in Eq. (4.61) may be set equal to unity.
The inequality of Eq. (4.62) always holds for long slits, then:

$$\frac{\Delta P}{H} \approx \frac{12\eta_\omega}{\delta^3} Q(t) \qquad (4.61\,a)$$

It should be kept in mind that two ultimate situations of filling through a short and long runner may be encountered in practice. Both cases were analysed in detail [36, 53, 54]. For the case of a short flat runner, the following expression for the coordinate of the flow front was obtained:

$$\frac{dh}{dt} \approx \frac{P_0\delta^2/(12\eta_\omega)}{h(t) + \frac{\eta_0}{\eta_\omega}\left(\frac{\delta}{\delta_0}\right)^3 h_0} \left\{ 1 - \frac{2P_0\delta^4\left(1 - \frac{8}{\pi^2}\right)\left(1 - \frac{T_0}{T_\omega}\right) E/RT_\omega}{\pi^4\varkappa\eta_\omega\left[h(t) + \frac{\eta_0}{\eta_\omega}\left(\frac{\delta}{\delta_0}\right)^3 h_0 \right]^2} \right\} \qquad (4.63)$$

The solution of Eq. (4.63) with the initial condition $h_0 = 0$ is too awkward, and here it will be omitted. But the analysis of this solution is simple and very important in practice. As compared to the isothermal case, the mould filling *slows down* for two reasons. First, this occurs due to an increase in the effective length of a channel, since in Eq. (4.63) the quantity $h_0(\eta_0/\eta_\omega)$ is present instead of h_0; this increase is essential because $(\eta_0/\eta_\omega) \gg 1$. Second, the mentioned slow-down in conditioned by the second term in the square brackets of the right side of Eq. (4.63) which is responsible for liquid heating upon its motion in a forming cavity. Note that in injection moulding at high pressure P_0 via a short runner, a strong dissipative heating of plastisol may take place within the runner, particularly when plastisol has enhanced viscosity $\eta_\infty(T)$.

For the case of a *long flat runner*, the equation for the filling front $h(t)$ has the form [36]:

$$\frac{dh}{dt} = \frac{P_0\delta^2/(12\eta_\omega)}{h(t) + h_0(\delta/\delta_0)^3}$$

$$\times \left\{ 1 - \frac{2P_0\delta^4(\delta/\delta_0)^2\left(1 - \frac{8}{\pi^2}\right)(1 - T_0/T_\omega) ER/T_\omega}{\pi^4\varkappa\eta_\omega[h(t) + h_0(\delta/\delta_0)^3]^2} \right\} \qquad (4.64)$$

$$h(0) = 0$$

The qualitative conclusions drawn from Eq. (4.64) are similar to those from Eq. (4.63). Note that in both cases the terms in the square brackets of Eqs. (4.63) and (4.64) in excess to unity are already small at t = 0 and do not exceed 0.15. These terms very soon become negligibly small due to fast growth of h(t). These considerations justify the applicability of the data of the isothermic calculations in Sect. 4.1–4.3 in engineering practice, privided that η is replaced by η_∞.

5 Analysis of the Final Stages of Injection Moulding of Plastisols

5.1 Pressure Equalization upon Mould Filling

Upon mould filling with plastisol, the device should be kept under pressure for some time, "feeding-up" a cavity with material to equalize pressure. This is very important for many reasons: it enhances shot density, homogenization of plastisol, it also prevents shrinkage and distorsion of the moulding, and as a result leads to essential improvement of article quality (enchanced strength, homogeneity, high precision of dimensions and shape).

At this stage, one has to take into account the volume compressibility of the material, since upon "feed-up" the "hold-on" time of material under pressure is determined by compressibility and slow viscous flow. If the pressure of injection P_0 is sufficiently high, then at this stage a liquid may be considered to be Newtonian with viscosity η_∞. Keeping this in mind, we may state that the calculation given below will be applicable to various plastisols (of types I and II) with the only difference that for plastisol I η = const, while for plastisol II η = η_∞. For the sake of simplicity, the analysis will be performed for the case of a flat mould filled through a slit runner (Fig. 10a).

To describe compressibility phenomena, let us make use of the equation of continuity averaged over the cross section of a flat cavity:

$$\frac{\partial P}{\partial t} + \varrho_0 \frac{\langle v \rangle}{\partial z} = 0 \qquad \langle v \rangle = -\frac{\delta^2}{12\eta} \frac{\partial P}{\partial z} \tag{5.1}$$

where $\langle v \rangle$ = Q/δ is specific consumption, P is pressure, z the longtitudinal coordinate, δ the slit width, ϱ_0 the density of the material in the "unperturbed" state (corresponding to atmospheric pressure P_0). Using the equation of the plastisol volumetric state in the linearized representation (where K is pressure modulus), from Eq. (5.1) we obtain:

$$\frac{\partial P}{\partial t} = \lambda \frac{\partial^2 P}{\partial z^2} \left(\lambda = \frac{K\delta^2}{12\eta} \right) \tag{5.2}$$

The so-called equation of "piezoconductivity" of Eq. (5.2) is widely used for the solution of similar problems, the quantity λ is usually called the piezoconductivity constant.

The initial condition for Eq. (5.1) corresponds to linear pressure distribution at the moment of complete mould filling when $h(t_f) = H$, and accordingly to Eq. (4.5) has the form:

$$P|_{t=0} = P_0 \frac{H - z}{H + h_0(\delta/\delta_0)^3} \tag{5.3}$$

The boundary condition at $z = H$ means impenetrability of the back wall of a mould, i.e. $\langle v \rangle|_{z=H} = 0$; taking into account the second equation of Eq. (5.1), it may be represented as $\dfrac{\partial P}{\partial z}\bigg|_{z=H} = 0$.

The boundary condition at the entrance into a forming cavity $(z = 0)$ may be derived from the following correlations[1]:

$$Q(t) \approx -\frac{\delta^3}{12\eta} \frac{\partial P}{\partial z}\bigg|_{z=0} ; \qquad Q(t) = \frac{\delta_0^3}{12\eta} \frac{P_0 - P|_{z=0}}{h_0} \tag{5.4}$$

Eliminating $Q(t)$ from Eq. (5.4), the boundary condition at $z = 0$ for pressure in Eq. (5.2) is readily obtained:

$$\left[P - \alpha H \frac{\partial P}{\partial z}\right]_{z=0} = P_0 ; \qquad \left[\alpha = \left(\frac{\delta}{\delta_0}\right)^3 \frac{h_0}{H}\right] \tag{5.5}$$

Now the complete formulation of Eq. (5.2) may be written in the form:

$$\left.\begin{array}{l}
\dfrac{\partial P}{\partial t} = \lambda \dfrac{\partial^2 P}{\partial z^2} ; \qquad P|_{t=0} = \dfrac{P_0}{1+\alpha}\left(1 - \dfrac{z}{H}\right); \\[3mm]
\left[P - \alpha H \dfrac{\partial P}{\partial z}\right]_{z=0} = P_0 ; \qquad \dfrac{\partial P}{\partial z}\bigg|_{z=H} = 0 ; \\[3mm]
\left[\lambda = \dfrac{K\delta^2}{12\eta}; \alpha = \left(\dfrac{\delta}{\delta_0}\right)^3 \dfrac{h_0}{H}\right]
\end{array}\right\} \tag{5.6}$$

The solution to the problem may be obtained by using the standard method of separation of variables and is given by:

$$\frac{P_0(z, t)}{P_0} = 1 - \frac{2}{1+\alpha} \sum_{n=-\infty}^{+\infty} \frac{\exp\left(-\dfrac{Cn^2\lambda}{H^2} t\right) \cos\left[Cn(1 - z/H)\right]}{Cn^2[1 + (\sin 2Cn)/(2Cn)]} \tag{5.7}$$

---

[1] The approximate character of the first correlation of Eq. (5.4) is due to neglection of the effects arising in going from one cross section (runner) to another (cavity) which are small for plastisols.

Equation (5.14) for the feed-up ($t_{h.o.}$) time do not contain injection pressure, as was the case in Eq. (5.9), for the time of pressure equalization with gelatination was not taken into account. Comparing Eqs. (4.14) and (5.12) reveals that the right side of Eq. (5.12) is by two orders of magnitude greater than the right side of Eq. (5.14), since $P_0 \sim 1$ atm and prssure modulus $K \sim 1000$ atm. Therefore it may be assumed that $t_{h.o.} \ll t_f$ and in the further analysis one way consider the temperature dependence of t_f as the limiting factor for the total period of the entire cycle, i.e. the productivity of moulding.

5.3 Analysis of the Temperature Dependence for Filling Time Under Conditions of Gelatination

Having determined t_f from Eq. (5.12) and making use of Eqs. (3.8)–(3.10) we obtain:

$$t_f = \frac{b_0}{a_0} e^{-(K+m)\Delta T} \left[\exp\left(S \cdot e^{m\Delta T}\right) - 1 \right] \tag{5.15}$$

where $a_0 = a(T_0)$; $b_0 = b(T_0)$; $\Delta T = T_\omega - T_0$; $S = \dfrac{6a}{P_0} \left(\dfrac{H}{\delta}\right)^2 \left[1 + 2\dfrac{h_0}{H}\left(\dfrac{\delta}{\delta_0}\right)^3\right]$.

Denoting $(S \cdot e^{m\Delta T}) = \xi$, let's rewrite Eq. (5.15) in the form:

$$t_f = \frac{b_0}{a_0} S^{(1+K/m)} \xi^{-(1+K/m)} (e^\xi - 1) \tag{5.16}$$

At $\xi \to 0$ and $\xi \to \infty$ for Eq. (5.16) we obtain the following asymptotic expressions [36]:

$$\left.\begin{array}{ll} t_f \approx \dfrac{b_0}{a_0} S^{(1+K/m)} \xi^{-K/m} & (\xi \to 0) \\[4mm] t_f \approx \dfrac{b_0}{a_0} S^{(1+K/m)} \dfrac{e^\xi}{\xi^{(1+K/m)}} & (\xi \to \infty) \end{array}\right\} \tag{5.17}$$

From Eq. (5.17) it follows that $t_f \to \infty$ at $\xi \to 0$ and $\xi \to \infty$. Thus we come to the very important conclusion from a practical point of view: the function $t_f(\xi)$ *has a minimum* at some $\xi = \xi_*$ corresponding to $\Delta T = \Delta T_*$. This minimum has a clear physical significance. Indeed, in the region of low temperatures (here $\Delta T < 0$) gelatination is suppressed, but viscosity $\eta(\tau)$ is relatively high (due to the normal Arrhenius dependence of viscosity on temperature). In the region of high temperatures (where $\Delta T > 0$), in spite of the decrease in the $b(T)$ component of viscosity, it strongly increases due to structuring ($a(T)$ sharply grows). Therefore, there should exist such η (and respectively the temperature $\Delta T = \Delta T_*$) that the filling time is minimal.

Determinations of the ΔT_* value is of great importance in practice. It was shown [36] that to ΔT_* and minimal t_f some certain quantity ξ_* may be matched up:

$$\xi_* = (1 + K/m)(1 - e^{-\xi_*}) \tag{5.18}$$

From Eq. (5.18) it follows that there exist the unique positive value of $\xi_* = \xi_*(K/m)$ that may be determined numerically as a solution to Eq. (5.18). When $\xi = \xi(K/m)$ is known, the value of ΔT_* may be determined by using $\xi = Se^{m \Delta T}$. Since $\Delta T = T_\omega - T_0$ and T_0 (or T_0') are room temperatures, we have:

$$T_\omega^* = T_0 + \frac{1}{m} \ln \left(\frac{\xi_*}{S} \right)$$ (5.19)

From Eq. (5.19) it follows that the value of T_ω^*, corresponding to minimal filling time for the given plastisol, is determined by the only parameter S which in turn is dependent on the injection pressure P_0 and geometric factors (parameters a_0, b_0, m, K are constants for individual plastisol at a given temperature T_0).

From Eq. (5.19) and the expression for S in Eq. (5.15) it follows that the optimal mould temperature T_ω^* may be raised with increasing injection pressure P_0 and decreasing geometric factor for a flat mould in the expression for S in Eq. (5.15). This factor is minimal at $\delta = \delta_* = (H/h_0)^{1/3} \cdot \delta_0$, and the value of S_{min} is given by:

$$S_{min} = \frac{18 a_0}{P_0} \frac{H^{4/3} \cdot h_0^{2/3}}{\delta_0^2} = S_*$$ (5.20)

Then:

$$(T_\omega^*) = (T_\omega)_{max} = T_0 + \frac{1}{m} \ln (\xi_* / S_*)$$ (5.21)

The analysis shows that the minimal filling time at maximal temperature of a hot mould T_ω is attained at maximal acceptable injection pressure P_0. Since the value of T_ω is also limited by safety considerations (thermal decomposition), the optimal value of injection pressure must be chosen on the basis of this limitation on the value of T_ω.

6 Design Technique for the Process of Mould Filling and Gelatination of Plastisols. Comparison of Theory with Experiment

6.1 Design Technique for the Process of Filling Cold-walled Moulds

In this section, the filling of thin and long cavities with geometric parameters satisfying the inequalities of Eq. (4.24) will be taken as a example. As may be seen, the condition $\varrho_0 < \delta$ in Eq. (4.24) is not essential, and hence the sign of inequality may be changed by the opposite one.

For the case of long cavities, $B \ll H$. In such a case, we may confine ourselves to the second, slowest filling stage, and complete information on the filling process is then provided by Eqs. (4.35)–(4.39). The mathematical model may be represented [36, 54] as:

$$\frac{h(t)}{B} = \sqrt{\left(\frac{2}{3\pi} h_0 \frac{\delta^3}{\delta_0^4} + \frac{\pi}{8}\right)^2 + \frac{P_0 \delta^2}{6\eta B^2} \Delta T} - \frac{2}{3\pi} h_0 \frac{\delta^3}{\varrho_0^4} \tag{6.1}$$

$$\Delta t = t - t_f'; \qquad t_f' = \frac{6\eta}{P_0}\left(\frac{B}{\delta}\right)^2 \cdot \frac{1}{4}\left[\ln\left(\frac{B}{2\varrho_0}\right) - \frac{2}{3\pi}\frac{\delta^3 h_0}{\varrho_0^4} - \frac{1}{2}\right] \tag{6.2}$$

$$\langle P(x,t)\rangle = \begin{cases} P_0 \dfrac{h(t) - x}{h(t) + B \dfrac{2}{3\pi}\dfrac{h_0}{\varrho_0}\left(\dfrac{\delta}{\varrho_0}\right)^3}, & h(t) \geq x \\[4mm] 0, & h(t) \leq x \end{cases} \tag{6.3}$$

For practical purposes, it seems to be convenient to represent Eqs. (6.1)–(6.3) by introducing the following dimensionless variables:
— characteristic time

$$t^0 = \frac{6\eta}{P_0}\left(\frac{B}{\delta}\right)^2 \tag{6.4}$$

— dimensionless variables

$$\tau = t/t_0; \qquad \hat{x} = x/B; \qquad \hat{h} = h/B; \qquad \hat{P} = \langle P\rangle/P_0 \tag{6.5}$$

— dimensionless parameters

$$\hat{I} = \frac{2}{3\pi}\frac{h_0}{\varrho_0}\left(\frac{\delta}{\varrho_0}\right)^3; \qquad \tau_* = \frac{1}{4}\left[\ln\left(\frac{B}{2\varrho_0}\right) + \hat{I} - \frac{1}{2}\right] \tag{6.6}$$

In such a case, Eqs. (6.1)–(6.3) become more compact and convenient:

$$\hat{h}(\tau) = \sqrt{(1 + \pi/8)^2 + \tau - \tau_*} - \hat{I} \qquad (\tau \geq \tau_*) \tag{6.7}$$

$$\hat{P}(\hat{x}, \tau) = \begin{cases} \dfrac{\hat{h}(\tau) - \hat{x}}{\hat{h}(\tau) + \hat{b}}, & \hat{h}(\tau) \geq \hat{x} \\[4mm] 0, & \hat{h}(\tau) < \hat{x} \end{cases} \tag{6.8}$$

As mentioned above, Eqs. (6.7) and (6.8) correspond to the second stage of filling when $\hat{h} \geq \hat{h}_* = \hat{h}(\tau_*) = \dfrac{\pi}{8}$. In this case, if the value of $\hat{x}$ is small, Eq. (6.8) will be applicable at sufficiently large times when $\tau \geq \tau_*$, in the opposite case pressure will vary across the slit width.

The dimensionless parameters $\hat{I}$ and τ_* in Eq. (6.6) are the measure of runner resistance and filling time for the initial ("semi-radial") filling stage, respectively.

Injection pressure P_0, mould temperature T_∞ (and the respective viscosity $\eta(T)$, and the thickness of the article δ are the parameters that could be varied in injection moulding.

Equations (6.7) and (6.8) do not contain P_0 and η (and thus they describe the generalized dependence on these parameters). As for δ, it appears in Î, and consequently Eqs. (6.7) and (6.) are not invariant with respect to ξ. But if Î ≪ π/8, which is the case when runner resistance is rather small (for plastisols this holds for the most practically important problems), parameter Î in Eqs. (6.6)–(6.8) will be omitted, and the formulae take the following simple appearence:

$$\hat{h}(\tau) \approx \sqrt{\left(\frac{\pi}{8}\right)^2 + \tau - \tau_*}, \qquad \tau \geq \tau_* \approx \frac{1}{4}\left(\ln\frac{B}{2\varrho_0} - \frac{1}{2}\right) \tag{6.7a}$$

$$\hat{P}(\hat{x}, \tau) = \begin{cases} 1 - \dfrac{\hat{x}}{\hat{h}(\tau)} & (\hat{h} \geq \hat{x}) \\ 0 & (\hat{h} < x) \end{cases} \tag{6.8a}$$

which is invariant with respect to P_0, η, δ.

Dimensionless filling time $\tau = \tau_m$ complies with $\hat{h} = \hat{H} = H/B$ and is given, according to Eqs. (6.7) and (6.6a), by:

$$\tau_m = \tau_* + (\hat{H} + \hat{I})^2 - (\pi/8 + \hat{I})^2 \tag{6.9}$$

$$\tau_m = \tau_* + \hat{H}^2 - (\pi/8)^2 \qquad \left(\hat{I} \ll \frac{\pi}{8}\right) \tag{6.9a}$$

Finally, in calculating the pressure at the running point x (coordinate) within the forming cavity, the dimensionless time τ_x nesessary for the x point to be reached by the flow front may turn out to be of use (i.e. the time x is needed for pressure to become non-zero at the x point). According to Eqs. (6.7) and (6.7a), this time is given by:

$$\tau_x = \tau_* + (\hat{x} + \hat{I})^2 - \left(\frac{\pi}{8} + \hat{I}\right)^2 \tag{6.10}$$

$$\tau_x \approx \hat{\tau}_* + \hat{x}^2 - (\pi/8)^2 \qquad (\hat{I} \ll \pi/8) \tag{6.10a}$$

Equations (6.4)–(6.10) are usually used in practical designing, the example of *detailed procedure* is given below[1]:

(1) Specify the geometry of a forming cavity (finished article) and a runner: H, B, δ, ϱ_0, h_0, and point x_i of pressure gauges (manometers) along the cavity.
(2) Specify the temperature and rheologic parameters (the temperature parameters of viscosity) of the given plastisol according to Eq. (3.6).
(3) Specify injection pressure (for instance, for very thin and long plates, $P_0' = 2$ atm and $P_0'' = 4$ atm).

---

[1] The examples of calculations down to numerical data and the comparison of theory with experiment see below and [36].

(4) By using Eqs. (6.5), (6.6), and (6.9), calculate the dimensionless parameters $\hat{I}$, $\hat{H}$, τ_*, τ_m, respectively.

(5) Compare $\hat{I}$ with $\pi/8$; if $\hat{I} \ll \pi/8$, the generalized (independent of δ) dependences will hold true.

(6) According to Eq. (6.4), calculate the characteristic time t_0.

(7) Calculate filling times from $t_m = \tau_m \cdot t_0$ (for each of the present values of ξ, P_0, and η).

To compare the prediction of theory with experiment, the time profiles for the pressures measured by pressure gauges at points x_i from the entrance into the cavity should also be calculated by using Eqs. $(6.7a)–(6.10a)^2$.

6.2 Design Technique for Mould Filling in Considering Plastisol Gelatination

In this case, the dependence of plastisol viscosity on mould temperature T_ω is given by the expressions of the type of Eqs. (3.8–3.10):

$$\eta = a(T_\omega) + b(T_\omega); \qquad a(T_\omega) = a_0(T_0) e^{m(T_\omega - T_0)};$$

$$b(T_\omega) = b_0(T_0) e^{-K(T_\omega - T_0)}$$

where the parameters T_0, $a_0 \equiv a(T_0)$, $b_0 \equiv b(T_0)$, m, and K are preliminarily determined in the rheological studies of the given composition.

As in Sect. 6.1, let us consider the injection moulding of a long thin plate through a slit runner, neglecting the resistance of the runner. Then Eqs. (4.12), (5.11), (5.12) for pressure distribution P(z, t), the motion of the averaged flow front h(t), and filling time will have the form:

$$P(z, t) \approx \begin{cases} P_0 \left(1 - \dfrac{z}{h(t)} \right), & z \leq h(t) \\ 0, & z > h(t) \end{cases} \tag{6.11}$$

$$\frac{6a}{P_0\delta^2} h^2(t) \approx \ln \left(\frac{a}{b} t + 1 \right) \tag{6.12}$$

$$\frac{6aH^2}{P_0\delta^2} \approx \ln \left(\frac{a}{b} t_f + 1 \right) \tag{6.13}$$

As before, the values of P_0 and δ are varied, and the mould temperature T_ω (with gelatination accounted) will be found from minimization for filling time t_f.

The respective expression for minimal t_f has the form of Eq. (5.16):

$$(t_f)_{min} = \frac{b_0}{a_0} S^{(1+K/m)} \xi_*^{-(1+K/m)} (e^{\xi_*} - 1)$$

---

^{2} In calculating the pressure at the point, at $\hat{\chi}_1 < h_*$ the time interval $\tau_* \leq \tau < \tau_m$ should be considered.

where ξ_* is the solution to the transcendental Eq. (5.18):

$$\xi_* = (1 + K/m)(1 - e^{-\xi_*})$$

and parameter S is determined by the moulding conditions of Eq. (5.15):

$$S = \frac{6a_0}{P_0}\left(\frac{H}{\delta}\right)^2$$

The optimal value of T_ω^* is found from Eq. (5.19):

$$T_\omega^* = T_0 + \frac{1}{m}\ln\left(\frac{\xi_*}{S}\right),$$

parameters $a(T_\omega^*)$ and $b(T_\omega^*)$ are given by the following expressions:

$$a(T_\omega^*) = \frac{a_0\xi_*}{S}; \qquad b(T_\omega^*) = b_0\left(\frac{S}{\xi_*}\right)^{K/m} \tag{6.14}$$

To make designing more convenient, let us introduce the characteristic filling time:

$$t_0 = \frac{b(T_\omega)}{a(T_\omega)} = \frac{a_0}{b_0}\left(\frac{S}{\xi_*}\right)^{1+K/m} \tag{6.15}$$

and dimensionless variables:

$$\tau = t/t_0, \quad \hat{h}(\tau) = h(t)/H, \quad \hat{z} = z/H; \quad \hat{P} = P/P_0 \tag{6.16}$$

Then Eqs. (6.11 and 6.12) may be simplified and take the form:

$$\hat{P}(\hat{z}, t) = \begin{cases} \dfrac{\hat{z}}{\sqrt{\dfrac{\ln(1+\tau)}{\xi_*}}}, & \tau \geq e^{\xi_* \hat{z}^2} - 1 \\[4mm] 1 - \sqrt{\dfrac{\ln(1+\tau)}{\xi_*}}, & \\[4mm] 0, & \tau < e^{\xi_* \hat{z}^2} - 1 \end{cases} \tag{6.11a}$$

$$\hat{h}(\tau) = \sqrt{\frac{\ln(1+\tau)}{\xi_*}} \tag{6.12a}$$

For the dimensionless time of filling τ_f we obtain:

$$\tau_f = \frac{t_f}{t_0} = e^{\xi_*} - 1 \tag{6.13a}$$

As is seen, Eqs. (6.17) and (6.15) lead to Eq. (5.16) for $(t_f)_{min}$. Note that the dimensionless expressions in Eqs. (6.11a)–(6.13a) are dependent only on the dimensionless parameter ξ_* which in turn is determined by the identity of plastisol, since, as follows from the transcendental Eq. (5.18), $\xi_* = \xi_*(K/m)$. Hence, Eqs. (6.11a) to (6.13a) are universal and applicable for any plastisol; consequently, experimental data should be represented for the variables of Eq. (6.16).

Design procedure:

(1) Specify the geometry of a forming cavity: H, δ, and the positions z_i of pressure gauges (from the beginning of a cavity).

(2) Specify the parameters of the given plastisol (from the data of rheological experiments): T_0, a_0, b_0, m, and K.

(3) Using Eq. (5.15), calculate the dimensionless parameter S for the given P_0 and δ.

(4) For a given plastisol, find the numerical solution ξ_* to Eq. (5.18).

(5) Determine filling time τ_f from Eq. (6.13a).

(6) From Eq. (5.19) determine the optimal mould temperature T_ω^* for the given P_0 and δ.

(7) Using Eq. (6.15), calculate the characteristic filling time t_0 for the given P_0 and δ.

(8) For the given P_0 and δ, calculate the practically important (optimal) filling time by using Eq. (6.13a).

(9) Determine the initial plastisol viscosity at optimal temperature T_ω and the given P_0 and δ: $\eta_0(T_\omega) = b(T_\omega)$.

(10) Estimate the ultimate plastisol viscosity $\eta(T_\omega, t_f) = a(T_\omega) t_f + b(T_\omega)$ at optimal temperature T_ω for the given P_0 and δ.

To compare theory and experiment, calculate pressure profiles at the points of pressure gauges (at the distance z_i from the entrance into a forming cavity). By using Eq. (6.11a) in the dimensionless representation, we obtain universal expressions:

$$\hat{P}_i(\tau) \equiv \hat{P}(\hat{z}_i, \tau) = \begin{cases} 0, & \hat{h}(\tau) < z_i \\ 1 - \dfrac{z_i}{\hat{h}(\tau)}, & z_i \le \hat{h}(\tau) \le 1 \end{cases} \tag{6.17}$$

$$\tau = e^{\xi_* h^2} - 1 \tag{6.18}$$

6.3 Comparison of Theory with Experiment

The theory of low-pressure injection moulding of plastisols and design procedures have been verified experimentally under consitions very close to those in industry [33-36, 54, 55] for the case of a flat thin mould for articles of plate type. The factory installation was additionally equipped with pressure gauges at several measuring points. The schematic diagram of the installation is presented in Fig. 12. There were one mould and two separable runners 80 mm long: around runner as a tubing 7 mm in diameter, which was considered as a "point" runner, and a cone flat tubing (shown separately in Fig. 12) flaring out from around entrance orifice 7 mm in diameter to the slit exit having 1.5 mm in width and 90 mm in length, which was equal to the width of the mould. This cone tubing was assumed to be a flat slit gat Y. The forming cavity had the dimensions $710 \times 60 \times 0.15$ mm (the slit width could be varied up to 1.5 mm with calibrated linings between half-moulds).

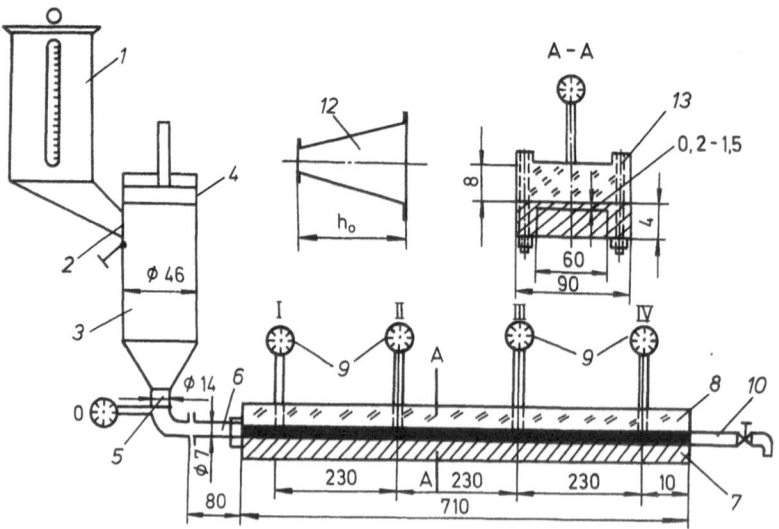

Fig. 12. Schematic diagram of the experimental set-up for plastisol moulding into thin flat moulds: 1 — loading of raw materials; 2 — damper; 3 — pump unit; 4 — plunger; 5 — discharge tube; 6 — inlet channel (round runner); 7 — metallic bottom half mould; 8 — top half mould (of lucite for low-temperature moulding and of steel for moulding into hot mould); 9 — pressure gauge; 10 — stub tube for control; 11 — drain cock; 12 — conical flat tube instead of round "point" runner; bolts for controlling the width of forming cavity within the limits of 0.15–1.5 mm

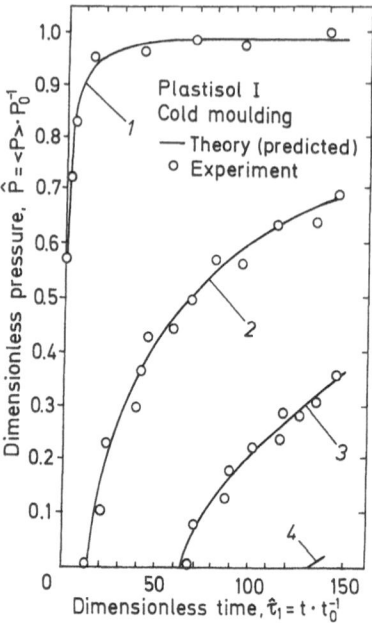

Fig. 13. Pressure as a function of time upon filling a cold mould with plastisol I. Both the parameters (P and $\hat{t}$) were calculated (solid lines) and experimentally determined (dotted) at four various points of the mould of Fig. 12, from 1 to 4, respectively

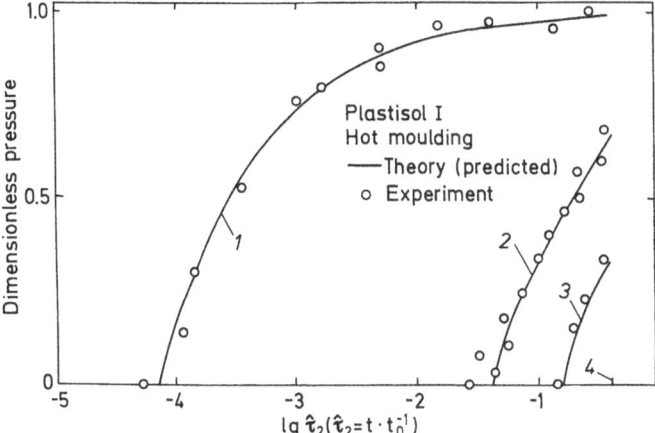

Fig. 14. Dimensionless pressure as a function of dimensionless time at four points of the mould for plastisol I with gelatination taken into account upon moulding into a preheated mould. Solid lines: calculated; dotted: measured

The lower half-mould was of stainless steel, the upper of two different materials: poly(methyl methacrylate) (lucite) 8 mm thick — in the case of low-temperature moulding without gelatination; and stainless steel — in the case of hot moulding. Four manometers were mounted at a distance of 230 mm from each other (manometers 1 and 4 were 10 mm from the mould edge). The manometers were calibrated against a high-frequency membrane pressure gauge of Daynisco Co. (USA). Pressures were measured with an accuracy of 2%.

In the experiments, the filling times for plastisols I and II were measured at various initial pressures and the values indicated by manometer: 1–4 were read as a function of time. The data obtained were represented in the dimensionless form as recommended in Sect. 6.3 and compared with the predictions of theory. The typical data

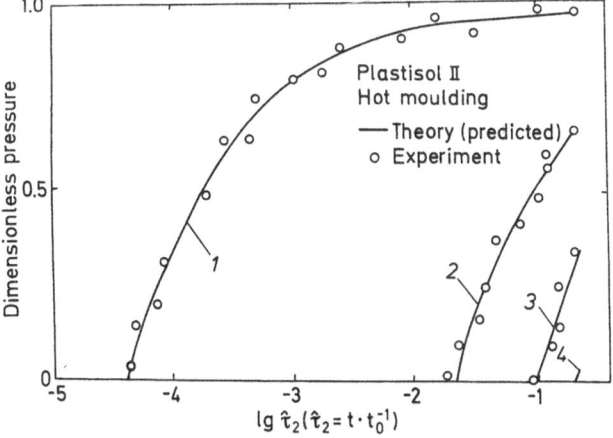

Fig. 15. Same as in Fig. 14 for plastisol II

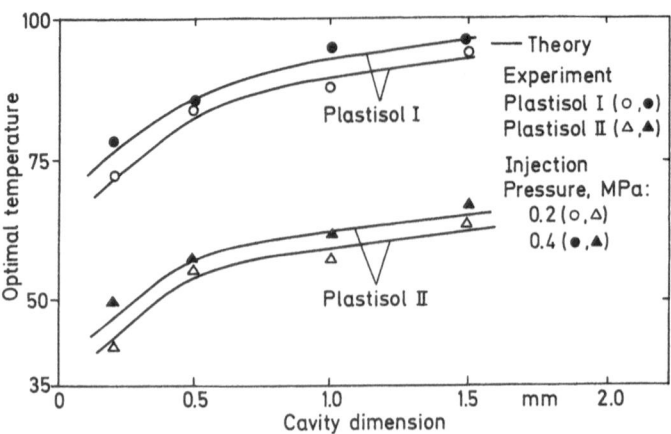

Fig. 16. Optimal mould temperature as a function of the inner cavity thickness for plastisols I (circles) and II

shown in Figs. 13–16 imply good agreement between theory and experiment. In spite of the encountered difficulties arising in view of the gelatination process, the uncertainty of the design is normally within 4–8%, reaching 10–12% only for some particular cases. Therefore, the suggested theory, mathematical models, and design procedures may be recommended for practical usage in the solution of a number of "direct" and "inverse" problems: the estimation of filling times and productivity, the determination of pressure profiles in a mould, the choice of the optimal temperature of a mould, the optimization of a cycle and parameters of moulding on the basis of minimal time, etc.

7 Low-pressure Moulding of Thermoplastics

Low-pressure processing of thermoplastic materials includes diverse technologies such as intrusion moulding, injection moulding of foam materials, manufacturing of long articles by mould casting. Nevertheless, detailed theoretical and experimental studies on the matter of interest are still lacking, and the design and development of equipment and accessories for low-pressure moulding is based mainly on intuition of development engineers. Below we shall analyse the most prominent contributions into the field of the last 2–3 decades.

It should be emphasized that in the case of thermoplastics, injection moulding will be referred to as low-pressure moulding for pressures not exceeding 5 MPa. This limitation provides a possibility to distinct confine the scope of our review and disregarding the processing by conventional automatic moulding machines.

According to Tadmor and Gogos [62], the process of mould filling may be partitioned into three stages, inherent to any kind of low-pressure moulding, and discussed in detail later. The following three stages of low-pressure moulding of thermoplastics will be discussed: (1) the entrance area ("gate"); (2) the area of developed flow of thermoplastic melt; and (3) the area in the vicinity of the flow front.

Let's proceed with the second case as the best studied area. In this case the analysis should be based on the general properties of flow of abnormally viscous liquids in tubing and channels under isothermal and non-isothermal conditions. Generally, the process of moulding may be considered as being isothermal when the following condition is satisfied:

$$\tau_f \ll \tau \approx \frac{L^2}{\alpha}$$

(7.1)

where τ_f if the filling time, τ the characteristic time of temperature change for an article formed, L the characteristic geometric parameters for the cross section of an article, and α the thermal diffusivity of a melt.

The melt flow under isothermal conditions, when it is described by the rheological equation for the Newtonian or power law liquid, has been studied in detail [63-66]. The flow of the non-Newtonian liquid in the channels of non-round cross section for the liquid obeying the Sutterby equation have also been studied [67]. In particular, the flow in the channels of rectangular and trigonal cross section was studied. In the analysis of the non-isothermal flow, attention should be paid to the analysis [68] of pseudo-plastic Bingham media.

In designing the accessories and choosing an injection and extrusion installation for low-pressure moulding[1], it is important to evaluate the design procedure for determining the channel resistance to non-Newtonian liquid flow. Practically simple and convenient design procedures for pressure losses and hydraulic resistance of non-round channels for power liquids and liquids described by the three-parameter model may be found, e.g. in [68,69].

The isothermal description, of course, is not applicable to various processes of low-pressure moulding, including manufacturing long-shaped articles by the method of low-pressure moulding using a screw extruder as plastificator. In this case, Eq. (7.1) is not satisfied, and for the adequate description of the process one has to take into account not only the motion of the material but also the thermal processes in the course of flow. The extreme complexity of the problem and the small number of publications devoted to the problem should also be noted. The analysis of individual technologies is practically absent in the literature, though some general features of non-isothermal flow of abnormally viscous liquids in tubings and channels have been studied by a number of workers. The general formulation of the problem and the first attempts of its solution were done in the classical works of Ballman, Shusman, Toor, Bird, et al. [70-73], who gave the set of non-linear differential equations in partial derivatives for the process under consideration. These authors carried out the first experimental studies on the melt motion in a channel at constant pressure [71]. They drew attention to the important role of material heating due to dissipation of energy at viscous flow and its cooling due to "thermal" expansion of a system. However, the large number of simplifying assumptions, including the neglection of the temperature dependence of viscosity, left room for further investigations in the field. The most complete solution to the problem, with energy dissipation, temperature

---

[1] Low-pressure moulding (filling moulds in making long shaped articles) may be performed using screw extruders as sources of thermoplastic melt.

dependence of the liquid, and flow parameters taken into account, was suggested by the Russian authors [74-77]. The criteria, providing a possibility to evaluate the conditions at which the wall temperature changes only slightly, so that in the analysis one may confine itself to the boundary conditions of the first kind, have been suggested in [74]. Communication [77] deals with materials exerting a sudden change in yield strength upon temperature increase and shows that there exist several flow regimes simultaneously: the area of the Newtonian flow; of quasi-solid flow, and of viscous shear flow. But it should be emphasized that the unwidely mathematics used in [74-78] could hardly be used for the analysis of individual processing procedures. The importance of hydrodynamic resistance in designing processing tools have already been mentioned above. For the case of non-isothermal flow, such a design procedure was proposed for media described by the three-parameter rheological model [74]. The non-isothermal flow in tubings for any given viscosity dependence on temperature has been studied [79] by performing qualitative analysis of integral equations. Some critical pressure gradient was found to exist, above which the stationary flow became impossible. The non-isothermal flow of the Bingham media has also been analyzed [80,81].

One of the most interesting and fruitful approaches to the theoretical study of the non-isothermal flow of polymer melts has been suggested by Merzhanov and co-workers [82-84], which reduced the set of equations for the flow to the well-known equation of thermal explosion. The authors obtained the profiles of temperature and velocity for the case of stationary flow. However, the thermal explosion was found [85] to become feasible at too high velocities $v \geq 10$ m/s which could hardly be attained at low-pressure injection moulding of polymers.

It should be emphasized again that all the cited papers were too general to be applied to the individual processing procedures. In our opinion, the approach utilizing linearization of the set of equations and its solution based on the principles of quantum-mechanical theory of perturbation seems to be most fruitful in analyzing the low-pressure moulding of long (up to 3 m) articles. Below we shall present our recent data obtained during the last two years.

Recall that in the process of low-pressure moulding, thermoplastic polymers (polyolefins, polyamides, or their compositions) are loaded into a cylinder of the adiabatic extruder, plasticized and injected into a mould at low pressure. The formed article is cooled in the mould and removed due to shrinkage phenomena.

For further analysis, let us choose the cylindrical system of references, the z axis of which is directed as shown in Fig. 17. Neglecting the dissipation processes, we obtain the following set of equations for the filling process:

$$\frac{\partial v}{\partial r} = \frac{r}{2\eta} \frac{\partial P}{\partial z} \tag{7.2a}$$

$$\frac{\partial T}{\partial t} + v(r, z, t) \frac{\partial T}{\partial z} = \frac{\varkappa}{r} \left(\frac{\partial T}{\partial r} \right) + \varkappa \frac{\partial^2 T}{\partial r^2} \tag{7.2b}$$

with boundary conditions:

$$T|_{t=0} = T|_{z=0} = T_0$$

$$T|_{r=R} = T_\omega$$

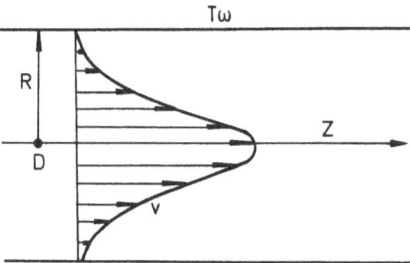

Fig. 17. System of references for the analysis of thermoplastic flow and the velocity profile across a mould

The temperature dependence of viscosity η is given by the conventional Arrhenius equation:

$$\eta = \eta_0\, e^{E/K\,T(r,\,z,\,t)} \tag{7.3}$$

where v is the velocity of melt motion, R the radius of a mould, T_ω the temperature of walls, $\varkappa$ thermal diffusivity, E the activation energy for viscous flow, and $\partial P/\partial z$ the pressure gradient.

The solution procedure was as follows: Linearize Eq. (7.2b) by assuming $v = \mathrm{const} = Q/\pi R^2$. Here Q is volume productivity of the plasticizing unit which is constant during the entire filling process and determined by its plastication parameters. Then find a solution to Eq. (7.2b) by the well-known methods of mathematical physics, and substitute it into Eq. (7.1a). Linearization of Eq. (7.2a) is performed by the expansion of $L'/KT(r, t)$ into the Taylor series. From Eq. (7.2a) and condition $Q = \mathrm{const}$, we find the expression for $v(r)$:

$$v_{(r)} = \frac{Q}{\pi l^2}\, e^{-(r/e)^2} \tag{7.4}$$

plotted in Fig. 17. Here $l = \dfrac{2}{3}\sqrt{\dfrac{KT^0}{U}}\,\dfrac{T_0 - T_\omega}{T_0}\,R$ has a clear physical significance as a characteristic path length at which flow velocity is changed remarkably.

Energy dissipation may be estimated by using Equation (5) from [63].

The lack of experimental data impose difficulties for modelling the processes of low-pressure moulding of thermoplastics. From this point of view, it is of interest to refer to [85] containing a wide scope of experimental material. The role played by energy dissipation as applied to flow in capillaries of viscosimeters was studied in [86]. To check the predictions of theory and to elucidate the applicability of one or another plastication unit, we have measured the pressure dynamics in the course of mould filling. Theory gives the following expression for pressure as a function of time at the head of an extrusion plasticator:

$$P = \frac{2Q\eta v}{\pi l^2 \varkappa}\left[1\frac{4\varkappa t}{l^2} - 1\frac{2\varkappa t}{l^2}\right] \tag{7.5}$$

In the experiments, pressure was measured with pressure gauge of the plunger type which transmitted pressure to calibrated manometers of high precision. During the first

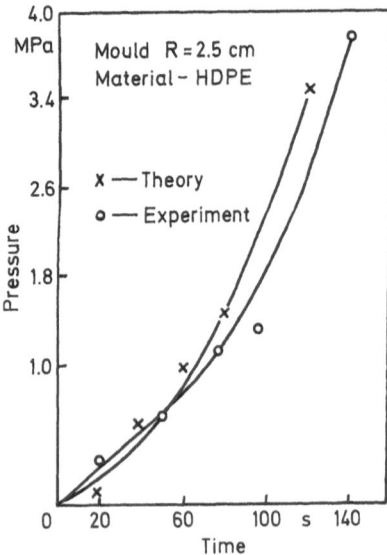

Fig. 18. Calculated and experimentally determined curves of pressure dynamics P(t) upon filling with R = 2.5 cm

several minutes of mould filling, good agreement between theory and experiment was found. The disagreement which follows may be evidently explained by considerable dissipation of energy during viscous flow.

In designing it seems very important to know about the absence of pressure at the mould wall in the course of filling. The pressure at the wall was experimentally found to be nearly zero due to the formation of a quickly solidifying skin of polymer in the vicinity of the wall.

Closing the discussion of the area of developed flow, note that a number of ideas concerning the modelling of low-pressure injection moulding may be borrowed from the studies of normal injection moulding which are reviewed in [36] and [62].

Now let us consider the third area in the vicinity of the flow front. This area draws the attention of the workers in the field of injection moulding, since this stage conditions the orientational phenomena and the properties of finished articles. The basic ideas concerning the behaviour and the character of the flow front in injection moulding were discussed by Fridman and Petrosyan [33-36] and by Tadmor and Gogos [62]. There is no doubt that the ideas are similar for both cases of low- and high-pressure injection moulding. Thus, the jet flow widely discussed in the literature [36,62,87] has been observed by us [88] in the study of low-pressure moulding in glass tubes instead of metallic moulds. The experimental determination of velocity profiles for the flow of the Non-Newtonian liquid from the data on the flow rate measurements in several tubings of small diameter at the given pressure gradient [88] may also be of considerable interest.

More distinct differences (as compared to high-pressure moulding) are encountered in the analysis of the first, least studied area of the entrance ("gate") into a mould. The authors of [89-93] subdivide the entrance area into the stages of jet and regular flow especially pronounced in moulding bulky articles. The jet flow has been visually observed by us in experiments with transparent moulds. The various stages of the jet flow, the process of "jet stacking", and the stage of a dense plate have been studied both theoretically and experimentally [91–93]. The authors state that the first difference in normal stress is almost the decisive factor determining the type of flow at the stage of jet flow. Thus, in filling a mould with LDPE showing a large difference of normal stress (mould radius 2 cm, nozzle diameter 1.4 cm) jet flow is not observed at all, and the flow is regular. In filling the same mould (and the same nozzle diameter) with HDPE showing much a smaller difference of normal stress, distinctly pronounced jet flow was observed. Of interest is the qualitative analysis of the set of equations for the flow front [94]. At any flow index (index of a power in the power law of flow), the velocity profile was found to have extrema, but extrema occur at various points for different flow indices; the profile is steeper for the dilate liquids, whereas it is more smooth for the pseudo-plastic liquid as compared to that of the Newtonian.

Note that some particular interest of technologists is paid to the processes of low-pressure moulding as applied to making multilayer long articles from thermoplastic blends. By controlling the parameters of a process, some ordering of components may be achieved, most often, for coaxial configuration.

The processes of foam thermoplastics moulding [95] may also be classified as the processes of low-pressure injection moulding. Plastication of material is achieved in the conventional screw unit. The proportioned quantity of melt of the volume smaller than that of a mould is loaded into a side assumulating cylinder with a plunger. The plunger presses material thus preventing its foaming. Then melt is admitted into a mould, plastic filling only the fraction of forming cavity.

Subsequently a mould is filled only due to the evolution of gaseous products of foaming agent ("porophore") decomposition. Their pressure, as a rule, has a value of 0.5–2 MPa. Sometimes, the screw itself is used instead of the accumulating cylinder. Just upon injection, pressure drops, and the spontaneous foaming of a composition thus begins. It may also be explained by oversaturating upon pressure decrease, due to which the great number of gas-filled bubbles is instantly formed which begin to grow. The system becomes to be separated into two phases, melt and gas. Novel moulding machines for this processing are described in various patents [96, 97].

The process of mould filling should not be considered as completely understood and controlled [98], though some important features of gas-filled system flow were established by Fridman and co-workers [99, 100] who have explained the formation of laminated structures in low-pressure moulding of foaming melts and found the correlation between moulding conditions and morphologic macro-structure and properties of foam articles.

It should be noted that the processes of low-pressure injection moulding are of importance not only in processing fresh polymeric materials of high quality, but for secondary row material thermoplastics as well, since the processes studied do not require melt filtration and hence power-consuming and technologically complicated stages of washing out, separation and drying of regrind materials.

In conclusion it may be noted that in spite of serious theoretical and experimental studies carried out in the field, the problem is still far from clearly understood. Let us formulate the most important questions to be solved. The first is the complete formulation and solution of the problem for mould filling either for the general case, or for individual processing procedures (aiming at shortening a moulding cycle, lowering power consumption, etc.) in mutual interdependence between the stages of the process. It seems also to be urgent to develop designing procedures for the accessories, estimating the feasibility of extrusion and moulding installation for the processing. The process of moulding under conditions of constant productivity (not constant pressure) has been studied insufficiently. The experimental data on the dynamics of mould filling, structuring in the course of low-pressure injection moulding, and the influence of processing conditions on the quality of finished articles are still lacking in the literature. All the problems may be successfully solved by using modern computational techniques.

Acknowledgements: The authors wish to thank Prof. A. I. Leonov for his help in the establishing of the theoretical part of this work, Dr. A. N. Prokunin for his valuable assistance in studying the rheological properties of plastisols and helpful discussions of mathematical modelling, and Dr. V. A. Teleshov who has provided the possibility of experimental checking of theoretical predictions.

8 References

1. Balakirskaya VL (1974) Polymeric pastes. In: Encyclopedia of Polymers. Moscow, Sovetskaya encyclopedia, vol 2, p 541
2. Masenko LY (1983) Study on PVC plastisol processing. Candidate's Thesis, Moscow, MITKhT
3. Andrianova GP (ed) (1981) Chemistry and processing of polymeric films and artifical leather. Legkaya i pistchevaya promyshl., Moscow, vol 1, p 262
4. Andrianova GP (ed) (1981) Chemistry and processing of polymeric films and artifical leather. Legkaya i pistchevaya promyshl., Moscow, vol 2, p 375
5. Shtarkman BP (1975) Plastification of polyvinyl chloride. Khimiya, Moscow, p 245
6. Hillen OR (1970) Gummi, Asbest, Kunststoffe 23: 932
7. Masenko LYa, Kagan DF, Shapiro GI (1980) Dipping of PVC plastisols. Express inf. "Khimich. promyshl." Ser. "Processing of plastics". NIITEKhIM, Moscow, 4: 6; 5: 7
8. Balabkin PI (1960) Manufacturing by dipping, Goskhimizdat, Moscow, p 265
9. Weber G (1969) Kunststoffe 59: 644
10. Stevens P (1970) Kunststoffe 60: 453
11. Levich VG (1959) Physics and chemistry of hidrodynamics. Fizmatgiz, Moscow, p 370
12. Deryagin BV, Levi SM (1959) Physics and chemistry of thin films coating onto moving support. Acad. Sci. Publ., Moscow, p 164
13. Gelperin NI, Nosov GA, Makotkin LV (1971) Teoret. Osnovy Khimich. Technologii 5: 429
14. Rubashkin BL (1968) Khimich. Mashinostr. 3: 24
15. Perepelkin KE (1978) Physics and chemistry of chemical fibers production. Khimiya, Moscow, p 114
16. Masenko LY (1981) Some features of PVC plastisols preparation and their coating onto polymer support. In: "Pererabotka i primenenie plastmass". Trudy NPO Plastik. NIITEKhIM, Moscow, p 37
17. Masenko LY, Kagan DF, Shapiro GI (1983) Extrusion of PVC plastisols. Plastich. Massy 5: 34
18. Baramboim NK (1961) Mechanochemistry of polymers, Rostekhizdat, Moscow p 409
19. Butyagin PYu (1971 B) Visokomol Soed. 13 (6): 428
20. Casale A, Porter RS (1979) Polymer Stress Reactions, vol 2 Academic Press, NY
21. Petrosyan AZ, Fridman ML (1985) In: Abstracts I National conf. on utilizing of recoverable polymers. Kishinev, vol 2 p 149
22. Masenko LY, Khandros VI, Panova TG (1975) Preparation of PVC plastisols in high-speed vacuum mixers. In: "Pererabotka i primenenie plastmass". Trudy NPO Plastik. NIITEKhIM, Moscow, p 92

23. Masenko LY, Shapiro GI, Bykhov VN (1976) An assembly for continuous production of reinforced polymeric hoses. USSR Inventors Certificate 699749
24. Masenko LY (1980) The method of plastisols extrusion. USSR Inventors Certificate 823160
25. Fridman ML (1988) Advances in equipment for mixing thermoplastic melts. TsINTINEPhTEKhimmash, Moscow, p 68
26. Tadmor Z (1979) US Pat 4 142 805; (1980) 4 194 841
27. Hold P, Tadmor Z, Valsamis L (1981) US Pat 4 255 059
28. Hold P, Tadmor Z (1982) US Pat 4 329 065
29. Tadmor Z, Hold P, Valsamis L (1979) A novel polymer-processing machine. Theory and experimental results. SPE 37th ANTEC, New Orleans, May 1979
30. Duran O (1980) Plast Technol Adv. 5: 216
31. Bykov AS (1973) Production of PVC Linoleum. Vysshaya Shkola, Moscow
32. Zhitinkin AA (1970) In: Khimiya i tekhnologiya vspenennykh polimerov", VNIISS, Vladimir, p 39
33. Fridman ML, Petrosyan AZ, Prokunin AN (1986) Manuscript deposited in VINITI 748-85 Dep., p 35 Bibl. ukazatel VINITI 1: 162
34. Fridman ML, Petrosyan AZ, Kazaryan GA (1986) Plast Massy 6: 16
35. Prokunin AN, Fridman ML, Petrosyan AZ (1986) Zhurn Prikl Teoret Fiziki AN BSSR 51: 609
36. Petrosyan AZ (1987) Mechanics of plastisols injection moulding. Candidate's Thesis, Moscow, Inst. Chem Phys AN SSSR
37. Subert W, Steller A (1967) Plaste und Kautschuk 1: 36
38. Zubchuk VA, Polyakova KK, Pichugina GA, Gubenskii VA (1971) Lakokrasochnye Materialy i ikh Primenenie 6: 74
39. Kroit GR (1955) Science of colloids. Inlitizdat, Moscow, vol 1 p 406
40. Severs ET (1969) Polymer rheology. Translation from English Malkin AY (ed) Khimiya, Moscow
41. Encyclopedia of Polymer Science and Technology (1971) New York vol 14 p 407
42. Kunststoff-Taschenbuch (1965) Karl Hanser, Munich vol 16 p 569
43. Scott R (1984) Brit Pat 2 135 324
44. Balakirskaya VL, Starkman BP (1969) Plast Massy 8: 54
45. Soldatov VM, Kirillov AI, Mol'kov AD (1970) Plast Massy 6: 5
46. Voronkova IA (1969) Investigation of paste-forming properties of PVC. Candidate's Thesis, Moscow, MIIKhT
47. Gudier (1964) PVC-rigisouls. In: Torner RV (ed) Pererabotka Polymerov, Translation from English. Chimia, Moscow, p 394
48. Hüls-Kunststoffe (1963) Brochure, Hüls, FRG, p 1
49. Starkman BP, Mukhina IA (1969) Plast Massy 5: 50
50. Rotenberg IP, Zapol'skaya KI, Sokolovskaya GP, Bobrik PP (1968) Plast Massy 1: 8
51. Koshlev VK, Gorshkov VS, Shapiro TM et al. (1976) Plast Massy 8: 53
52. The Weissenberg Rheogoniometre Instruction Manual for Model R18 (1977) Sangamo Controls Ltd., Bogner Regis, England, p 47
53. Leonov AI, Basov NI, Kazankov YV (1977) Fundamentals of pressure moulding for thermosetting plastics and rubbers. Khimiya, Moscow, p 216
54. Fridman ML, Petrosyan AZ, Prokunin AN (1986) Manuscript deposited in VINITI 749-85 Dep., 43 pp Bibl Ukazatel VINITI 1: 162
55. Petrosyan AZ, Movsesyan EA, Fridman ML, Prokunin AN, Kazaryan GA (1986) Prom Armenii 7: 38
56. Custro JM, Lipschitz SD, Masosco CW (1982) AIChE Journal 28: 973
57. Schmitt LR (1974) Polym Eng Sci 14: 797
58. Domine JD, Gogos CG (1980) Polym. Eng Sci 20: 847
59. Malkin AY, Sherysheva LI, Kulichikhin SG, Zhirkov FV (1983) Polym Eng Sci 23: 804
60. Malkin AY, Zhirkov PV, Berezovskii AV (1984) Polym Proc Eng 2: 207
61. Godovskii YuK (1982) Teplofizika polymerov. Khimiya, Moscow 293 pp
62. Tadmor Z, Gogos CG (1979) Principles of polymer processing John Wiley, New York
63. Landau LD, Lufshitz EM (1986) Hydrodynamics, Nauka, Moscow, p 79

64. McKelvy DM (1965) Pererabotka Polymerov. Translation from English. Khimiya, Mosocow, 442 pp
65. Vinogradov GV, Malkin AY (1977) Rheology of polymers. Khimiya, Moscow, 433 pp
66. Astarita D, Marucchy G (1979) Osnovy gidrodinamici Nenjutonovskikh gidkosty. Translation from English. Mir pabl., Moscow, 309 pp
67. Mitsuishi N, Aoyagi I (1969) Engng Sci 24: 309
68. Pure and Appl Geophysics (1970) 83: 82
69. Devisilov VA, Belov SV (1979) Izvestiya Vys Uchebn Zaved. Mashinostroenie 12: 50
70. Lestchii NP (1986) Hidrovlika i Hidrodinamika 43: 7
71. Ballman RL, Shusman T, Toor HL (1959) Mod Plast, Oct: 115
72. Toor HL (1955) Trans Soc of Rheology 11: 35
73. Bird RB (1957) SPE Journal 1: 177
74. Tyabin NV, Dakhin OKh, Baranov AV, Gerasimenko VA (1982) Teplofizika Vysokikh Temperatur 1: 81
75. Dakhin OKh, Baranov AV, Tyabin NV, Gerasimenko VA (1983) Teplofizika Vysokikh Temperatur 4: 740
76. Tyabin NV, Dakhin Okh, Baranov AV, Gerasimenko VA (1983) Ingenerno-fizicheskii Zhurnal 45: 380
77. Benderskaya SL, Khusid BM, Shulman ZP (1980) Mekhanika Zhidkosti i Gaza 3: 3
78. Radionova NV, Froishtetter TB, Danilevich SYu (1980) Promyshl Teplotekhnika 2 (4): 51
79. Kachanov SA (1962) Zhurn Prikl Mekhaniki i tekhn Fiziki 3: 82
80. Mamedov RM, Sattarov RM (1977) Mekhanika Zhidkosti i Gaza 2: 162
81. Lomanov AS (1977) Zhurn Prikl Mekhaniki i tekhn Fiziki 6: 86
82. Bostandjiyan SA, Merzhanov AG, Khudyaev SI (1965) Doklady AN SSSR 163: 28
83. Bostandjiyan SA, Merzhanov AG, Khudyaev SI (1965) Zhurn Prikl Mekhaniki i Tekhn Fiziki 5: 45
84. Merzhanov AG, Stolin AM (1971) Doklady AN SSSR 198: 1291
85. Radionova NV, Froishtettez TB (1979) Fundamentals of chemical engineering, Theoret Osnovy Khimich Tekhnol 13: 226
86. Gerard SE, Steidler FE (1965) Ind Chem Fundam 4: 332
87. Tadmor Z (1974) J Appl Polym Sci 18: 1753
88. Kann KB, Feklistov VN (1980) Inzhenerno-fizicheskii Zhurnal 42: 927
89. Basov NI, Kazankov YuV (1984) Injection moulding of polymers, Khimiya, Moscow, 307 pp
90. Basov NI, Kazankov YuV, Lyubartovich VA (1980) Design technique for manufacturing and processing of polymeric materials, Khimiya, Moscow, 242 pp
91. Gorodnichev YuV, Basov NI, Kazankov YuV (1974) Proizvodstvo shin, RTI i ATI 7: 15
92. Klass A, Mueller M (1974) Kunststoffe 64: 135
93. Gorodnichev YuN, Basov NI, Kazankov YuV (1978) Proizvodstvo Shin, RTI i ATI 10: 10
94. Zhizhin TV (1986) Zhurn Prikl Mekhaniki i tekhn Fiziki 2: 69
95. Semerdjiev SG (1979) Thermoplastichnie konstruktsionnie penoplasty, Chem. Mash. Publisher, Sofia
96. US Pat (1984) 4 473 516
97. FRG Pat (1984) 3 346 418
98. Menges G, Schrammk (1973) Kunststoffe-Rundschau 20: 165
99. Sabsai OYu, Fridman ML, Nikolaeva NE (1984) Doklady AN SSSR 276: 140
100. Nikolaeva NE, Fridman ML, Sabsai OYu (1985) Rheological properties of gas-filled polymers as applied to moulding, NITEKhIM, Moscow, ISSN No 0203-7793

Editor: M. L. Fridman
Received July 18, 1988

Granulated Thermosetting Materials (Aminoplasts) — Technology

V. I. Tunkel and M. L. Fridman

USSR Institute of Plastic Materials Research, Perovsky proezd 35, Moscow, USSR, 111112

In the present survey an attempt is made to systematize and generalize data contained in earlier technical and patent publications concerned with production of granulated thermosetting materials (aminoplasts) and its development trends. Also presented are the classification of the existing granulation methods and facilities and a comparative analysis thereof. Description is made of the results obtained in studying various technological features in the production of granulated aminoplasts by pelletizing with the aid of a plate granulator.

Consideration is given to the most promising technological systems in granulation of aminoplasts and types of facilities used.

An outline is made of technical solutions providing for commercial production of high-quality granulated aminoplasts.

Advances in Polymer Science 93
© Springer-Verlag Berlin Heidelberg 1990

1 Introduction

At the present time aminoplasts comprise moulding compounds based on amine resins (thermosetting condensation products of formaldehyde with carbamide or melamine or a combination thereof) and organic or mineral fillers or a combination thereof doped with dyestuff and modifying substances (plasticizers, stabilizers, crosslinking agents, and the like).

Aminoplasts represent one of the most common types of polymeric materials suitable for manufacturing a wide range of general industrial, electrotechnical and domestic articles. Available in powdered, granulated and fibrous forms, aminoplasts are processed by compression, transfer and injection moulding.

Aminoplasts are characterized by good physico-mechanical and dielectric properties. They are light-resistant, odorless, readily colorable (capable of taking up various hues, including light shades), possess high surface hardness at normal and high temperatures, arc resistance and ability to quench the electric discharge due to liberation of gases (nitrogen, hydrogen, etc.). Aminoplast articles can withstand the effect of weak acids, alkalis, lubricating oil, alcohol, acetone, benzene, petrol and other solvents, as well as domestic chemicals.

Amine-resin thermosetting materials (plastics) are manufactured in all industrially developed countries. Aminoplasts are still one of the most common types of polymeric materials although production of novel plastics is rapidly growing.

Table 1 below gives figures characterizing production and consumption of aminoplasts in the leading industrial countries in recent years [1].

Table 2 lists the most commonly used grades of aminoplasts manufactured by firms of some of the leading industrial countries [2].

Several types of aminoplasts are manufactured in the USSR to serve different purposes [3]:

Table 1. Production and consumption of aminoplasts in leading industrial countries in 1981–1985

Country	Annual output, thousand tons				
	1981	1982	1983	1984	1985
FRG, total	—	70.0	73.5	74.5	74.9
Urea-formaldehyde materials	—	13.9	18.7	18.9	14.8
Great Britain	—	127.0[a]	125.0[a]	122.0[a]	125.0[a]
France	185.0	165.0	170.0	160.0	156.0
	229.0[a]	217.0[a]	198.0[a]	218.0[a]	200.0[a]
Japan, total	—	—	631.0	605.0	578.0
Urea-formaldehyde materials	—	—	511.0	488.0	470.0
Spain	—	96.0	89.5	106.0	101.0
	—	—	—	93.0[a]	88.0[a]
USA	—	512.0[b]	601.0[b]	629.0[b]	647.0[b]
Australia	—	—	—	40.0[a]	49.0[a]

[a] Consumption data; [b] Sales data

Table 2. Grades of aminoplasts manufactured by firms of some leading industrial countries

Country	Firm	Trademark	Characteristic of material
1	2	3	4
USA	Allied Chemicals	Plaskon Hg UJg	Based on urea-formalde-resin and cellulose filler; processable by moulding and injection moulding
		Plaskon MTg TWg MJg	Based on melamine-form-aldehyde resin and cellu-lose filler; processable by transfer, compression and injection moulding
	American Cyanamid	Bettle B W (1342)	Based on urea-formalde-hyde resin and cellulose filler; processable by moulding and injection moulding
USA	American Cyanamid	Cymel 10776 9481 107 T	Based on melamine-form-aldehyde resin and cellu-lose filler; processable by moulding and injection moulding
Great Britain	BIP Chemicals	Beetle	Based on urea-formalde-hyde resin and cellulose filler; processable by moulding and injection moulding
		Melmex	Based on melamine-form-aldehyde resin and cellulose filler
	JCJ	Mouldrite	Based on urea-formalde-hyde resin and cellulose filler
France	Prochal	Cibanoide	Based on urea-formalde-hyde resin and filler
		Melolam	Based on melamine-form-aldehyde rerin and filler
	Plastimer	Uralite C	Based on urea-formalde-hyde and cellulose filler
		Ervamine	Based on melamine-formaldehyde resin and cellulose filler
FRG	Hoechst	Hostaset VF	Based on urea-formalde-hyde resin and cellulose filler; processable by moulding and injection moulding
FRG	Hoechst	Hostaset MF	Based on melamine-form-aldehyde resin and cellulose filler; processable by moulding and injection moulding
	Dynamit Nobel	Pollopas	Based on urea-formalde-hyde resin and cellulose

Table 2. (continued)

1	2	3	4
			filler; processable by moulding and injection moulding
		Ultrapas HMz	Based on urea-melamine-formaldehyde resin and cellulose filler
		Ultrapas Mz	Based on melamine-form-aldehyde resin
	BASF	Urecolle F	Based on urea-formalde-hyde resin and cellulose filler
Italy	Montedison	Gabrite	Based on urea-formalde-hyde resin and cellulose filler
		Melbrite	Based on melamine-form-aldehyde resin and cellu-lose filler
	STA Staliano Resine	Sirit	Based on urea-formalde-hyde resin and cellulose filler
Italy	STA Staliano Resine	Melser	Based on melamine-form-aldehyde resin and cellu-lose filler
Switzerland	Ciba-Geigy	Cibadoine	Based on urea-formalde-hyde resin and cellulose filler
		Melopas	Based on melamine-urea resin and cellulose filler
	Studli	Bulitol 32a Zc	Based on urea-formalde-hyde resin and cellulose filler
		Bulitol 42a Mc	Based on melamine-formaldehyde resin and cellulose filler
Sweden	Perstorp AB	Skanopal 100-Series 300-Series	Based on urea-formalde-hyde resin and cellulose filler
		Isomil 700-Series	Based on melamine-form-aldehyde resin and cellulose filler
Japan		Leadlite	Based on urea-formalde-hyde resin and cellulose filler

— KΦA type (KΦA 1 and KΦA 2 grades) — a general-purpose aminoplast based on carbamide-formaldehyde resin and loaded with cellulose filler.

The KΦA 1 grade is used for making translucent industrial and household artic-les which are not intended to be in contact with food products (scales, wiring acces-sories).

The КФА 2 grade is used for making opaque industrial and household articles intended to be in contact with loose food products, as well as stationary, toys, haberdashery, cans, bottles and wiring accessories;

— МФБ type (МФБ 1 grade) — a melamine-formaldehyde resin aminoplast loaded with cellulose filler.

Articles made of the МФБ-type aminoplast have a hard, scratch-resistant surface and are generally stronger than articles made of the КФА-aminoplast. Heat stability on exposure to moisture, solvents and domestic chemicals and also good electric properties permit utilization of the МФБ-type aminoplast for making electrotechnical parts intended for operation at elevated humidity and temperature.

The aforementioned aminoplast is also suitable for making electrotechnical articles (external parts of electric shavers, etc.), articles intended to be in contact with food products (tableware for use in commercial flights, etc.);

— МФВ 1, 3 and МФВ 4 grades) — a melamine-formaldehyde resin aminoplast loaded with organic (cotton pulp) and mineral (asbestos, talc) fillers.

The МФВ 1 grade is used for making arc-resistant electrotechnical articles, including those for shaft equipment.

The МФВ 3 grade is used for making parts of vehicular ignition assemblies.

The МФВ 4 grade is used for injection moulding of electrotechnical articles;

— МФД type (МФД 1 grade) — a melamine-formaldehyde resin aminoplast loaded with filler (asbestos doped with a small amount of cotton pulp), processable by hot moulding.

The МФД 1 grade is used for making electrotechnical articles characterized by stringent arc-resistance and heat-stability requirements;

— МФЕ type (МФЕ 1 grade) — a melamine-formaldehyde resin and glass-fiber aminoplast possessing increased mechanical strength, heat- and arc-resistance and used for making electrotechnical articles characterized by stringent arc-resistance, heat-stability, mechanical-strength and wear-resistance requirements in normal and moist tropical wheather conditions (at a relative humidity of 98 % and a temperature of 35 °C). The aminoplast is processed by hot moulding.

The type КФА aminoplasts are in large-scale production in the USSR. However, only powdered materials habe been made until recently, a disadvantage increasing labor and material costs in subsequent processing operations.

The aforementioned disadvantage may be obviated by granulation, that is, by forming a given material into particles (pieces) of about the same size in a predetermined size range.

Granulated aminoplasts do not produce dust and may be readily stored and transported. Their bulk is stable, other distinguishing features being adequate looseness and increased thermal conductivity. The use of granulated aminoplasts permits automatic measuring of the material in moulding operations, the associated advantages being accurate and stable supply of processing machines, an increased mould closing speed, fewer material losses, increased output of processing equipment, creation of conditions for full automation of processing operations by the use of rotary lines, fully automatic presses and automatic thermosetting plastic forming devices with simultaneous improvement of shop sanitation conditions and decrease of fire hazard.

Whatever the processing method, aminoplasts intended for use in modern industrial equipment should satisfy the most stringent requirements to stability of technological properties whose variations hinder automation of a production process, for moulding conditions may have to be changed to suit a particular lot of the material.

The change-over to automated processing of thermosetting materials calls for utilization of novel equipment, improved processing technology and replacement of powdered materials with up-to-date granulated materials.

Increased demand for granulated thermosetting materials and their effective use necessitated the selection and utilization f the efficient methods known in the art or provision of new rational techniques for granulation of aminoplasts, development and introduction of highly productive equipment to meet the aforementioned requirements.

The subjects under discussion in the present survey have been chosen in view of the fact that pertinent periodicals generally lack any systematic data on the granulation theory, technology and equipment, as applicable to aminoplasts.

Inadequate information on the above subjects contrasts with detailed and comprehensive outlines of the theory of flow of processed materials presented by Soviet (Basov NI, Leonov AI, Kazankov Ya V et al.) and foreign (Darnell WH, Mol FA, Scheider K, Mondvai J et al.) scientists in periodic and monographic publications.

The principles of granulation of materials through agglomeration by pelletizing described in earlier publications apply, to our knowledge, primarily to fertilizers and metal powders and not to aminoplasts.

We believe that modern development and further improvement of aminoplast granulation technology are primarily dependent on the level of effective techniques designed and on the provision of refined production facilities.

For the above reasons, the authors concentrated on technological, engineering and designing aspects of the aminoplast granulation problem.

In the present survey an attempt is made to systematize and generalize data contained in technical and patent publications and relating to production of granulated materials based on urea-formaldehyde resins with cellulose filler. An outline is also made of the results of some of the investigations carried out by Soviet specialists in 1983–1986.

2 Granulation of Thermosetting Materials. Technology, Facilities and Development Trend

For about half a century, technological problems relating to the production of thermosetting materials have been in the center of attention of scientists in different countries. During this long period, different approaches have been taken and quite a variety of technical means and methods have been used to solve the above problems.

To select an optimal aminoplast granulation technology, it is expedient to analyze the related art with a view to tracing scientific and technological trends, as regards granulation of thermosetting plastics, and finding the most promising technical solutions.

By analogy between thermosetting plastics and thermoplastic materials whose granulation is well developed and brought to a commercial level in many countries, two main directions may be conventionally defined as follows [4]:
— granulation on completion of manufacturing process;
— granulation during production stage.

The latter direction may be a priori regarded as technically advantageous and more promising.

A major disadvantage of granulation on completion of th e manufacturing process is th e need to transfer a solid material to a plastic or liquid state, whereas in production of thermosetting plastics such a state is passed in homogenization, that is, the granulating operation substantially complicates the entire production process. Such a situation may be justified only if a portion of the material to be supplied to the user is to be granulated (but not the entire lot).

Both methods of granulating thermosetting plastics are being developed in the Soviet Union. However, it should be noted that the method involving granulation of thermosetting plastics on completion of the manufacturing process was devised first, which occurred in 1954.

As regards granulation of thermosetting plastics during the production stage, applications for USSR inventor's certificates were first filed in 1966.

Most of the inventor's certificates and patents in the priority countries relate to the method of granulation on completion of the manufacturing process. However, account should be taken of the fact that the method of granulation during the production stage came into practical use only in the mid sixties.

Still, comparing the total number of relevant national applications filed between 1963 and 1984 it may be concluded that most of them relate to the method of granulation of thermosetting materials during the production stage.

It i s obvious that the choice of a granulation method should be made considering the requirements for a given material and its processing conditions, which is a major factor in mastering novel granulation processes.

2.1 Granulation of Aminoplasts.
Technology, Facilities and Development Trend

Aminoplast granulation technology and facilities have not been hitherto satisfactorily studied and described in relevant scientific publications.

An analysis was made of the different methods of granulation of aminoplasts [5]; the paper contains a classification scheme of the methods presented in Fig. 1.

The study of different methods of granulation of aminoplasts shows that the given process involves compaction of an homogeneous powdered product with subsequent sizing and forming of the compact product.

Figure 2 is a schematic diagram illustrating the granulation of aminoplasts [6].

The preparation involves introduction of liquid and/or powdered additives into powdered basic material. Liquid additives may be introduced into separate tanks beforehand (periodically or continuously) or the material may be passed through a chamber in which the liquid is sprayed. Powdered additives may be introduced in periodic or continuous mixers.

The use of the following constituents is described in Ref. [5]: water, glycerol (or its solution), wax solution in water-soluble ketones, aliphatic (stearic) alcohol and ethanol mixtures, urea-formaldehyde solution, aqueous solutions of urea, vinacryl, impregnating urea-formaldehyde resin, polyoxyethylene, etc.

Compaction of the material is the main stage of the production process. The prepared

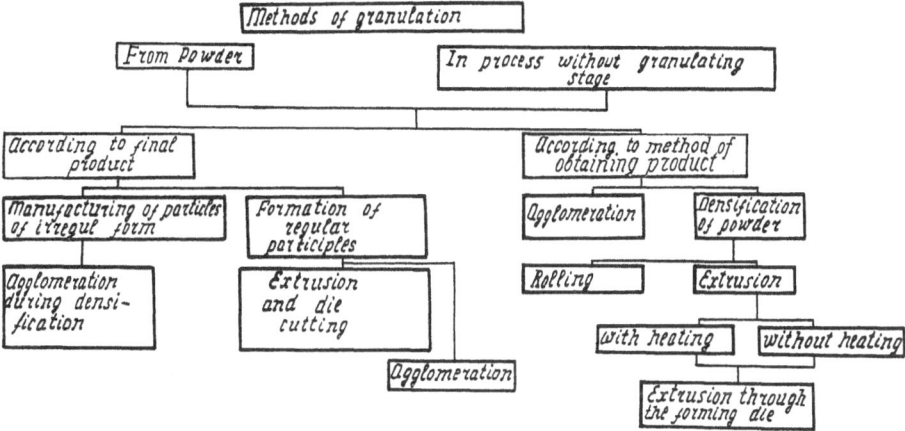

Fig. 1. Modern aminoplast granulation techniques (schematic diagram)

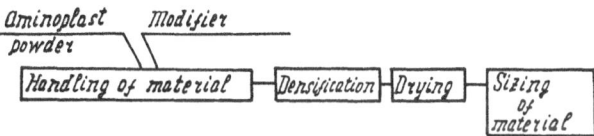

Fig. 2. Granulation of aminoplasts. Block Diagram

mixture may be compacted by different means such, for example, as high-speed mixers with anchor (impeller) agitators, horizontal screw mixers with rotary or oscillating motion of working members and a free output, rolls screw mixers with extrusion heads, etc.

Introduction of liquid additives increases the moisture content, which necessitates drying of the material.

It may be dried before or after moulding in fluidized-bed continuous-belt or rotary-tube driers.

The material is shaped and sized by milling the compact mass with impact mills or by forcing it through dies with subsequent cutting by special knives.

With reference to the proposed classification (Fig. 1) consideration will now be given to the most interesting technical solutions relating to methods and facilities for making granulated aminoplasts and found primarily in patent publications.

2.1.1 Granulation on Completion of Manufacturing Process

The method of obtaining granulated moulding materials according to a USSR Inventor's Certificate [7] comprises the step of mixing powdered aminoplast with the liquid constituent followed by agitations at a temperature of 70 to 110 °C. At an earlier stage, the powder is mixed with glycerol, loaded in a screw machine and agitated with the temperature in heating zones of the machine being reduced from 110 to 90 °C. At this temperature the material is softened, compacted and unloaded from the machine as spherical particles.

Another prior-art method of granulating aminoplasts according to a USSR Inventor's Certificate [8] comprises the steps of mixing powdered aminoplast with the liquid constituent (a 1 to 5% urea aqueous solution) by agglomeration into granules, their drying and subsequent screening. The obtained granules possess an increased curing rate and flow.

The so-called agglomerating liquid disclosed in a USSR Inventor's Certificate [9] provides for obtaining strong granules in pelletizing moist powdered aminoplast on a rotary plate of the granulator. The formed granules are supplied to a drying apparatus wherein excessive moisture is removed. Thereafter the granules are sized to obtain commercial fractions.

Disclosed in a Swiss Patent [10] is an installation fo r granulating powdered or pasty materials. The installation comprises a pair of oppositely rotating rollers disposed, respectively, on the inside and outside of a rotary cylindrical element to limit a gap wherein the material is compacted in holes provided in the cylindrical element. The material is forces through in a string-like manner onto the inner surface of the cylindrical element. Thereafter a subsequent pair of rollers extrudes the strings onto the outer surface of the cylindrical element, whence they are cut by a suitable knife arrangement.

Another prior-art method of granulating powdered aminoplast loaded with a cellulose filler is disclosed in a British Patent [11]. In such a method, granulation occurs in a container with compressed air being intermittently supplied through a hole, which is required to transfer the powder to a turbulent (cloud-like) state. A granulation liquid (for example, water supplied through a nozzle) is simultaneously sprayed in the container. Each air supply cycle lasts 0.5 to 5 s, the interval between pulse sequences being 0.5 to 10 s. After moistening, the powder agglutinates. The agglutination process may be expedited by requisite facilities. After drying with warm air, the granules are let out through a hole in the container.

At an earlier stage, the powders may be mixed with pigments, lubricant and other additives, which are fed into the container through injectors.

To improve quality and facilitate maintenance of the known installation, a roll mechanism according to a USSR Inventor's Certificate [12] comprises rolls displaced relative to one another in a vertical plane, while the proposed device for controlling the amount of material in a loading zone represents a special rack.

A known powder granulating installation according to a US Patent [13] includes a rotary drum in conjunction with a moistener sprinkling device. Provision is also made for preliminary moistening of the source powder.

The design of a Schugi mixer-granulator [14] ensures vigorous agitation of mixture constituents and a fairly high granulation rate, separate granules being characterized by adequate homogeneity. The main unit of the granulator is a mixing chamber with a high-speed shaft disposed centrally. The shaft mounts the mixing blades, which are not equidistant and have an individually adjusted slope. Powdered material containing an appropriate liquid (a binder solution) injected thereto is supplied from above to the mixing chamber. As the known apparatus passes the flow composed of air, a small amount of powder, and the liquid sprayed in the air, sinter is formed, the size of its particles being appropriately controlled.

2.1.2 Granulation during Production Stage

The device for granulating polymeric materials proposed in this country in 1966 and disclosed in a USSR Inventor's Certificate [15] is characterized by that its grinding mechanism comprises a rotor operating in a chamber and having elastic scrapers secured thereto. The ends of the scrapers completely shut off the flow of the supplied material as the rotor makes one revolution. The material is granulated in a soft state at the final stage in a mixer-homogenizer.

In accordance with a French Patent [16] granulation is accomplished in the course of mixing due to a combination of rotary and pulsating movements of the mixing elements.

Disclosed in a US Patent [17] is a continuous process of compacting powdered aminoplast providing for uniform density of the obtained granulated product. A particular aminoplast compound is dried, powdered and mixed with additives whereupon water is added so that the system may flow in compaction. The water is subsequently removed and the powder is then treated with steam for moistening and heating. Next, the compound is compressed and formed into a sheet by mixing rolls. The sheet is dried and finally ground.

Another known method of manufacturing granulated moulding materials based on aminoresins is desribed in an FRG Patent [18] whose prototype in Switzerland is the National Patent [19]. First, a dry mixture composed of fillers, pigments, and accelerators and previously homogenized is obtained in a high-capacity premixer. Thereafter lubricant is added and the mixture is agitated until desired homogeneity is obtained. Next, the mixture with an aminoresin solution is supplied to a slowly rotating mixer, granulated and dried. The foregoing method is advantageous in that plasticity of an homogeneous mixture is obtained due to an aminoresin solvent and not by heating, which precludes premature polycondensation of a thermosetting material occurring in heat treatment.

A known method according to an FRG Patent [20] granted to Werner and Pfeiderer involves continuous production and granulation of moulding materials. In this method, double-helical machines commonly used with thermoplastic materials are utilized for the first time with thermosetting polymers to ensure continuous granulation. The invention resides in that thermosetting materials are prepared and granulated in a screw machine without any intermediate stages whereby the obtained product is suitable for subsequent processing. After passing through the first mixing zone, in which it is melted, homogenized and partially condensed, the compound is doped with a constituent inhibiting the condensation reaction and decreasing the viscosity of the system. This additive, which may be water for some moulding materials, is supplied to the compound in the second mixing zone.

The equipment based on a double-helical machine also comprises a separable extrusion head with a die and a granulator. The screws of the cascade machine are arranged in succession. They turn counter-clockwise and are fitted in one another. Several holes provided in the frame of the machine accommodate flanges adapted to secure metering arrangements. The above holes are also suited to pass through the inhibitors. The screws may be provided with special nozzles to expedite the mixing process (sectional mixing elements). The compound is forced out of the mixing zone and passes through the perforated die. The obtained strings are then cut into granules.

A prior-art method of obtaining small-size dense granules and a device thereof according to a Swiss Patent [21] mark the first attempt to obtain granulated thermosetting plastics during the production stage.

However, difficulties involved in complying with the stringent accuracy requirements for operating conditions stipulated by the aforementioned patent caused the patent holders to waive their patent rights after a period of three years.

A known method of continuous granulation in a boiling (fluidized) layer and a device therefore are disclosed in a Japanese Patent [22].

An analysis of the patent situation relating to the problem of granulation of thermosetting materials shows that, for the past decade, no novel methods have been proposed in addition to the known techniques involving the use of rolls, screws and agglomeration by pelletizing.

Patended by firms in different countries are primarily the improvements relating to the widely known and commercially used production processes.

Moreover, a qualitative analysis of patent, scientific and technological information on the design of basic granulation facilities confirms that:

1. Investigations conducted in 1975–1985 with a view to designing basic production facilities for granulation of thermosetting plastics concern primarily with improvements relating to plate granulators, screw mixers and rolls.

2. In the early eighties, Buss A.G., which is one of the leading Swiss chemical concerns, has been most active firm in protecting its rights in Switzerland and abroad, as regards screw mixer-granulators (cf. US Pat. No. 4, 044, 788, PCT No. 8, 001, 148, EPB 0, 033, 351).

The firm's prime efforts are directed at improving the design of internal working assemblies to provide for an increase in the yield of desired fractions.

A wide range of mixers and granulators of standard sizes manufactured by the firm are advertised and offered for sale on the world market. Also known in the field are the West German Werner and Pfleiderer, Uniroyal AG., the U.S. Beloit Corporation and some of the Soviet firms.

Investigations relating to the design of mixer-granulators, plate granulators and roll mechanisms are currently conducted in the USSR and abroad by the West German Bayer AG and Eirich, the Dutch Schugi, and the Japanese Mitsui Toatsu Kagaku KK.

Protected by patients and certificates of authorship are several technological improvements, more specifically, changes in the ratios of powder to agglomerating liquid and in the composition thereof to control the granule size (agglomeration method) and improve the design of rolls with a view to obtaining optimal temperature conditions and increasing the effectiveness of the production process (roll method).

3 Principal Aminoplast Granulation Methods

Referring to the above information on granulation technology and equipment, it is generally possible to divide all prior-art methods into two major groups, depending on the granulation techniques used (1) methods involving compaction (plasticization) of powdered aminoplast (adding, if required, various modifiers) by the use of rolls, screw mixers and other machines with subsequent crushing and moulding of the compound with the aid of dies; (2) granulation methods associated with agglomeration by pelletizing (with powdered and agglomerating liquid introduced into corresponding equipment).

With the methods relating to the first group, powder is compacted in a closed volume or between adjacent planes. It is conventional practice to heat the powder. After drying (or without this stage), the obtained compound is crushed and sized or forced through calibrating holes in a requisite moulding tool.

In agglomeration, the powder is compacted during a mixing stage adding, as a rule, a predetermined amount of moistening substance. Particles in the powder stick together increasing in size. After drying, they may be used as a hard product or be crushed to a predetermined size.

Consideration will now be given to technical solutions characterizing the two granulation methods mentioned above.

3.1 Granulation Involving Compaction (Plasticization) of Materials

As has been already noted in the foregoing discussion, compaction is regarded to be the main stage in granulation of aminoplasts.

One of the most effective technical means for compacting the material is a screw set although its utilization in the aminoplast granulation process involves considerable difficulties. High viscosity and low plasticity of the processed aminoplast cause great resistance to flow in the dies and also dissipative heating of the material, which leads to premature solidification of the polymer. One of the suitable techniques decreasing the aforementioned undesirable effect involves introduction into the source aminoplast of liquid or powdered additives decreasing viscosity, increasing plasticity and hindering condensation.

The compaction difficulties necessitate special requirements of the design of main working members, more specifically, the screw which is one of the most important elements. Of great significance are also accuracy of maintaining predetermined temperature conditions and automation of different production stages.

In a recently published extensive survey [23] consideration is given to main trends relating to development and improvement of modern equipment for mixing and plasticizing thermoplastic materials. Discussion below will be restricted to facilities for thermosetting plastics.

In the Paudal mixer-granulators [24], the moistened compound is transferred, compressed and forced through a moulding plate.

Such a machine comprises a single- or double-screw externally heated extrusion granulator, which may be one of the following types, depending on the direction in which the material is forced through the moulding plate: longitudinal unit (Fig. 3) or transversal unit (Fig. 4) [24].

Figure 5 shows the design of the screws comprised in the longitudinal extrusion granulator. Figure 6 depicts the roller assembly of the transversal extrusion granulator.

The manufacturing firms resort to different technical solutions providing for effective mixing and homogenization of materials in the cylinder of the screw machine,

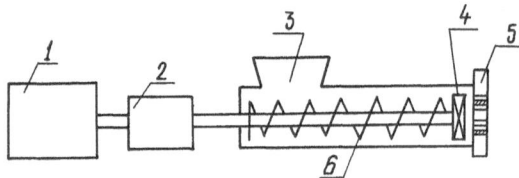

Fig. 3. Longitudinal extrusion granulator.
1 Motor; 2 reduction gear; 3 bin; 4 forcing blade; 5 nozzle; 6 screw.

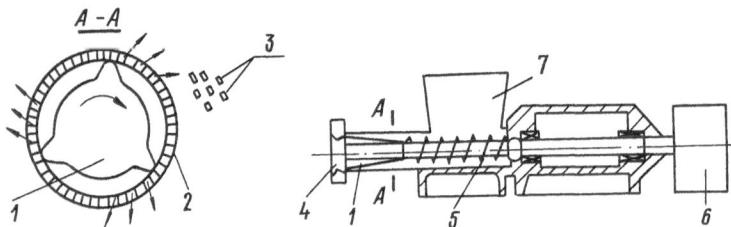

Fig. 4. Transversal extrusion granulator.
1 Forcing roller; 2 mesh; 3 granules; 4 locking screw; 5 screw; 6 motor; 7 bin.

a

b

c

Fig. 5a–c. Granulator screws.
a Uniform long pitch; **b** variable long pitch; **c** with conical core

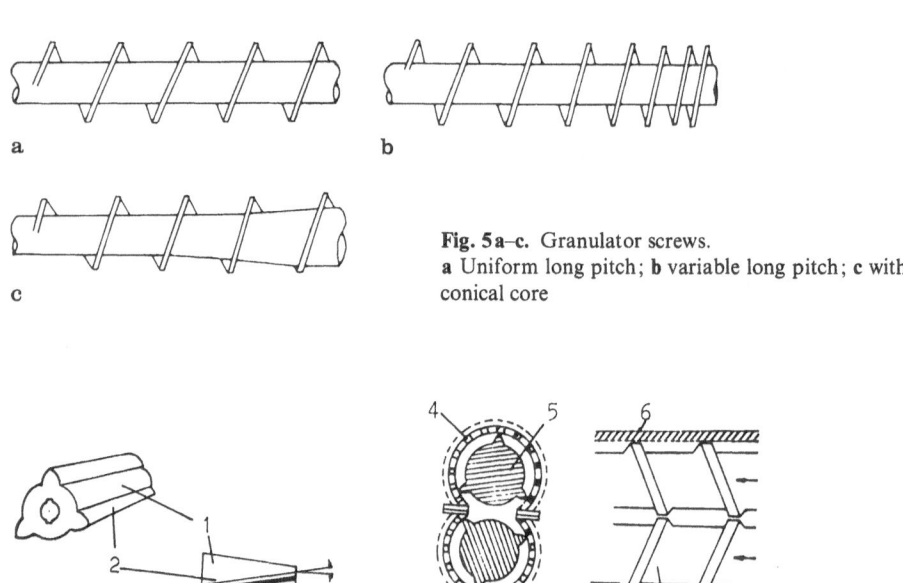

a

b

Fig. 6a, b. Forcing roller assembly (**a**) and top view of double screw (**b**).
1 Forcing roller; 2 forcing blade; 3 direction of feed; 4 mesh; 5 forcing roller; 6 screw enclosure; 7 screw feed

Fig. 7. Elements of screw for granulating thermosetting plastics

primary attention being given to the design of the screw which is the main unit of the machine.

The screw of standard design (Fig. 7) normally comprises a through shaft mounting in a predetermined succession different screw bushes (elements) and special kneading sections composed of discs turned relative to one another.

For example, the Berstorff extruder-granulators [25] incorporate a screw comprising different functional elements, more particularly, transfer elements with profiles inclined at different angles (Fig. 8), kneading and mixing elements (Fig. 9) providing for effective dispersion of additives and homogenization of materials. The kneading elements may be arranged individually or sectionally (successively in the unit). An additional mixing effect is obtained owing to a different width of discs (washers) or their left-hand or right-hand entry.

The mixing sections are used for distributing liquid, powdered and fibrous additives. These sections are composed of separate or several successively arranged toothed discs differing, as a rule, in their number, width and tooth form.

The screws of the Baker Perkins mixers [26] manufactured in the USA and Great Britain are also sectional units.

In double-screw mixers, the screws normally rotate in one direction at the same speed, an advantage ensuring a high rotational speed and increasing a shearing force and the output of the entire machine.

Furthermore, the screws rotating in one direction are free from additional (outward) loads increasing wear of the screws and the heating cylinder of the mixer.

With the screws rotating in one direction, the material is transferred from one screw to another so that the screws are constantly loaded, a feature facilitating formation of an homogeneous mixture and ensuring optimal conditions for heat exchange between the material and a casing wall. Also, degassing will occur more readily.

The Buss AG continuous mixers [27] comprise a special screw performing an oscillatory motion. The screw has three slots per lead, whereas kneading bolts or teeth in the mixer casing are aligned with the slots in the screw as the shaft rotates.

The shaft reciprocates axially as it completes a revolution. This mechanical motion

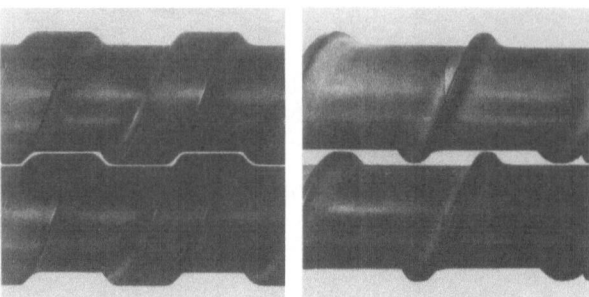

Fig. 8a, b. Screw transfer elements.
a Tight-fitting thread; **b** loose-fitting thread

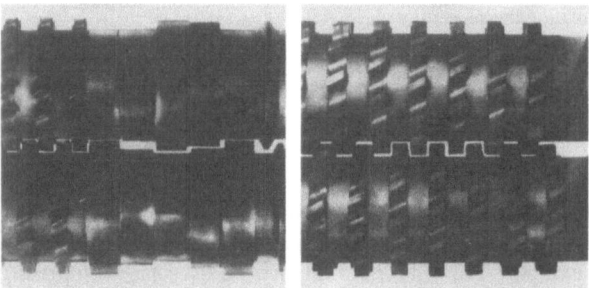

Fig. 9. Plasticizing (a) and mixing (b) screw elements

ocurring due to a special drive is rigidly coordinated in two planes. Combined axial and rotary motion of the shaft causes relative movement of the material between the slots and the kneading teeth. As a result, the edges of the slots in the screw first move past the teeth, the compound found in the formed harrow slot being suitable cut. The energy applied to the material caused a phase transition (melting) of the solid compound or laminar flow (interstratification). During combined reciprocating and rotary motion of the shaft, the slots in the screw enter the space between the kneading teeth. The compound found therein is, thus, axially displaced, thereby providing for the local mixing effect.

In addition, the laminar-layer flows formed in the so-called shear slots are subsequently disintegrated in the tooth space. This effect resembles interstratification of material between calender rolls.

As the motion cycle is completed, the slotted screw section returns to the original position.

A ZSK-type double-screw mixer proposed by Werner & Pfleiderer (Table 3) [28] adapted to manufacture solidified moulding compounds.

The mixer comprises two adjacent screws rotating in one direction at the same speed. The meshed screws provide for self-cleaning of the machine. The foregoing mixer incorporates sectional screws made up of screw bushes and mixing discs, which are circumferentially displayed relative to one another.

The heating cylinder of the mixer is normally made up of separate casing elements

Table 3. Specifications of Werner & Pfleiderer Double-Screw Mixers manufactured in FRG

Type of mixer	Screw dia, mm	Output, kg/h	Production length, max.	Drive power at screw speed of 300 I/min	Screw cutting depth, mm	Dimensions (1 × w × h), mm	Total mass, kg
ZSK25	25	35, max.	42	—	4.15	210 × 78 × 140	950[a]
ZSK30	30	35, max.	42	5.4	4.7	220 × 78 × 140	950[a]
ZSK40	40	60–150	48	—	7.1	260 × 72 × 140	850
ZSK53	53	100–350	48	32	5.5	380 × 55 × 130	3000
ZSK57	57	100–350	48	32	9.5	380 × 55 × 130	3000
ZSK58	58	100–350	48	32	10.3	340 × 70 × 140	3000
ZSK70	70	400–800	48	–	12.5	400 × 70 × 155	4800
ZSK83	83	500–1500	48	150	7.5	530 × 70 × 140	7000
ZSK90	90	500–1500	48	150	13.5	530 × 70 × 140	7000
ZSK92	92	500–1500	48	150	16.3	600 × 70 × 145	7300
ZSK120	120	1000–3000	48	4000	10.5	790 × 160 × 120	15,000
ZSK130	130	1000–3000	48	400	19.5	790 × 160 × 120	15,000
ZSK133	133	1000–3000	48	400	23.5	750 × 170 × 150	20,000

[a] Including drive control panel

(Fig. 10) mounted n a common foundation. The last two elements of the casing are made detachable for ease in cleaning the machine when the compound is solified.

All the elements of the casing have holes to allow its heating in respective zones or water cooling.

The Buss mixers are of two types: the PRD mixer serves for pre-mixing, homogenization and cold crushing of materials and the DKG machines are provided with a shaft and a granulating screw in an air-granulation cascade arrangement (Table 4)[27].

The casings of these mixers having a length of 4D to 23D are provided with one or

Fig. 10. Mixer casing elements

Table 4. Buss Screw Mixer-Extruders manufactured in Switzerland

PRD type	Mixer modification			
	46	100	140	200
DKG type	—	10–14	14–18	20–25
Aminoplast output, kg/h	5–15	120–150	270–350	600–800

more holes adapted to discharge gas or introduce the necessary constituents. A distinguishing feature of the mixer design is that its casing may be opened vertically to facilitate inspection and cleaning of the screw and the casing interior. The heating cylinder is provided with separable half-shells. The kneading sections may have holes suited to mount temperature sensors and inject liquid constituents.

The Berstorff mixers are made up of interconnected sections, each of which has a length of 5D (2.5D or 7.5D, if required).

Also, the cylinder casing elements have holes adapted to discharge gas and introduce liquid constituents. These elements are made integral or include shells of nitrated steel.

The temperature is maintained within desired limits by the use of oil, water or steam.

The Baker-Perkins double-screw mixer-extruders manufactured in the USA and Great Britain are employed for processing thermosetting materials, more specifically, urea-formaldehyde compounds. Basic specifications of these units are given in Table 5 [26].

A distinguishing feature of the Baker Perkins mixers is that their casings may be opened horizontally to permit inspection or cleaning of the screw or cylinder.

The mixer-extruders proposed by the aforementioned manufacturing firms boast of high automation, control and monitoring levels. This feature characterizes not only the mixer unit but also the metering systems, vacuum and granulation devices. Moreover, the control system makes it possible to optimize the production process by obtain-

Table 5. Bakter Perkins Double-Screw Mixer-Extruders manufactured in the USA and Great Britain. Basic Specifications

Specifications	Type, modification				
	MP2065TM	MP2080TM	MP2100TM	MP2125TM	MP2160TM
Motor power, kW	28	56	112	225	450
Output (of urea formalde-hyde compounds), kg/h	300	600	1000	1800	3500
Length, mm	3600	4450	5290	6600	8400
Width, mm	860	900	950	1400	1800
Height, mm	1800	1950	2300	2900	3200

ing, analyzing and correcting basic parametric data such, for example, as stock temperature, pressure, rotational speed, output and the like.

The granulating technology involving the use of screw mixers have not hitherto found wide applications in the USSR due to lack of high-quality facilities in the related field.

3.2 Granulation Involving Agglomeration By Pelletizing

The mixer-granulators may be different machines, the most common of which are:
— plate granulators;
 · eddy mills;
— fluidized-layer granulators.

The plate granulators ensure continuous granulation. Powdered aminoplast is supplied to an inclined rotary plate whereby it is lifted and then lowered. Agglomerating liquid is forced into the formed layer by injectors. The powder moistened with the liquid is pelletized. The formed pellets reaching a predetermined size are fed to the plate and then by gravity to an area wherein drying and subsequent sizing occur.

The plate granulators are manufactured both in the USSR and abroad (for example, by the West German Eirich company) [29].

The Eirich granulators may be of different standard sizes (Table 6). The specifications of the Soviet-made granulators are given in Table 7 [30].

The Schugi granulator [14] is an example of an eddy mill. A similar machine is designed in the USSR. The principal assembly of such a granulator is a mixing chamber with a centrally disposed high-speed shaft having blades. The slope of each blade can be adjusted. Powdered aminoplast is loaded into the upper part of the chamber and a suitable liquid (binder solution) is injected thereto. Under the action of the high-speed blades, the powder and binder flows become turbulent, thereby providing for uniform and rapid distribution of the binder within the powdered compound, which results in rapid growth ($\sim$ 5.0 s) of granules. As the formed granules move striking against the casing wall and the blades, compaction occurs and the granules are unloaded for subsequent processing.

Table 6. Specifications of Eirich Plate Granulators manufactured in the FRG

Specifications	Model						
	TR10	TR14	TR22	TR29	TR36	TR45	TR60
Plate diameter, mm	1000	1400	2200	2900	3600	4500	6000
Plate height, mm	280–500	400–700	600–925			800–1100	600–800
Plate drive, kW	1.1–2.2	3.7–5.5	11–22	22–45	37–45˙	44–55	75–110
Slope relative to horizontal, deg.	30–92	50–92	50–92	50–92	50–92	50–92	40–75
Mass, kg	600	1700	3700	7000	10,000	14,000	26,000

Table 7. Specifications of Soviet-made plate granulators

Specifications	Model					
	0T100	0T160	0T250	0T300	0T350	0T550
Plate diameter, mm	1000	1600	2500	3000	3500	5500
Plate height, mm	130–260	200–400	300–620	350–750	400–800	500–1100
Drive power, kW	1.3	3.5	10.0	16.0	23.0	55.0
Output, m³/h	0.2	1.5	8.0	13.0	19.0	45.0
Mass, t	0.6	1.8	4.0	5.8	7.5	19.8

The product stuck on the casing in granulation is removed due to reciprocating motion of the rotor. The granulometric composition of the product may be regulated by changing the speed of the granulator shaft, the slope and number of the mixing blades and by selecting suitable liquid additives, the latter technique being of prime importance.

The granulators developed in the USSR have a chamber diameter of 150 to 400 mm and an output of 0.3 to 25 t/h.

In the event of continuous and intermittent fluidized-layer granulators [31], the source powder is mixed due to creation of a "boiling" (fluidized) layer in the air flow, the powder being sprinkled with a suitable liquid until granules are formed.

3.3 Principles of Granulation by Pelletizing

One of the earlier publications [32] contains a detailed description of general principles of granulation by pelletizing with a plate granulator and also an analysis of the granulation mechanism and kinetics.

Lets now consider the fundamentals having theoretical and practical importance, as applicable to the aminoplast granulation technology and equipment.

3.3.1 Granule Formation Mechanism

The procedure of granulation by pelletizing includes four major stages:
— mixing of source powder with binder;
— formation of granules from small particles and crushing of lumps;
— pelletizing and compaction of granules moving over the surface of the plate;
— stabilization of the granule structure due to stronger links after a liquid-solid transition.

At all the stages, the particles are distributed according to size, the process being dependent on the techniques and granulation facilities utilized and also on the properties of the product.

3.3.2 Mixing and Granule Formation Stages

Granulation by pelletizing involves preliminary formation of agglomerates from uniformly moistened particles or deposition of dry particles on moistened substance in the granulation centers. Such a process is attributable to the action of capillary adsorptive forces between particles and to subsequent compaction of the structure due to forces of interaction between the particles within the compact dynamic layer.

The capillary effect within the loose material layer is dependent on the amount of a binder liquid at the contact point, on the contact form and on the number of contacts per unit volume.

The maximum size of the formed lumps is directly proportional to the size of a binder drop and inversely proportional to porosity of the material layer. The liquid will no longer propagate in loose material as soon as a given lump reaches the maximum capillary moisture capacity.

The powder is moistened and compacted simultaneously under the action of capillary forces. The smaller the surface tension, the more compact is the agglomerate.

3.3.3 Pelletizing Stage

Compaction of particles primarily occurs as they strike from different directions against the stationary material layer or the granulator sides. The value of kinetic energy imparted to a given lump depends both on its falling speed and mass. As the lumps repeatedly roll down and make impacts, compaction occurs and excessive moisture rises to the surface, thus attracting dry particles. As the particles are drawn together, the thickness of the bound water films decreases, while their adhesion grows. In operation of the granulator, the granules may be crushed (due to looser links between the particles), stuck together (when the liquid phase is in excess), or disintegrated (when the amount of liquid is insufficient), depending on kinetic peculiarities of the moisture absorption process. With the above factors taken into account, it is normally advantageous that the powder should be moistened repeatedly to obtain granules of desired size.

The efficiency of the granulator is directly dependent on the filling factor of the pelletizing surface, i.e. the granulator plate. Optimal conditions are obtained when small fractions are separated from the granulator side at the top of the plate, while larger particles roll down before the separation. The maximum rolling speed should not exceed the speed at which the granules are disintegrated. This speed depends on the properties of the granulated material and is determined experimentally.

Knowing the tolling speed it is possible to determine the diameter and the angle of inclination of the plate.

With varying values of the filling factor and the plate speed, the falling flow of the material has a different thickness at a constant rolling speed, mobility of particles in the given layer being decreased with an increasing amount of the rolling material. Consequently, the maximum output of a desired fraction may be obtained only at a predetermined filling factor of the plate.

The angular speed of the plate is dependent, in turn, on its diameter and angle of inclination. This speed is chosen to ensure separation of particles at the top of the plate.

The amount of material on the plate at a predetermined angle of its inclination should provide for optimal pelletizing conditions and also for a required time of its stay on the plate. Knowing the above characteristics and the output it is possible to determine the height of the granulator side.

3.3.4 Granule Structure Stability Stage

A liquid phase is removed to strengthen links between particles compacted in pelletizing, which ensure plasticity of the material and make it possible to change the shape of granules without disintegration. Drying is one of the most widely used methods of strengthening granules.

3.4 Principal Factors Affecting Granule Formation in Pelletizing

The density and size of granules are primarily determined by such design features of the granulator as the plate diameter, side height, and the angle of inclination of the plate and also by operating conditions of the granulator (plate filling factor and speed, time of stay of material on the plate).

However, granulometric composition is also dependent on several technological factors, more specifically, the properties of source material and binder liquid, their ratio and particle size of material on the plate.

The granulation process was investigated using a Soviet-made granulator having the plate diameter $D = 1800$ mm and side height $L = 270$ mm. The plate speeds were 0.81, 1.12, and 1.39 s^{-1}. The output (referred to source powdered aminoplast) was 150 kg/h. The angle of inclination of the plate was varied within 50–60°.

One of the main problems generally facing the investigators is the choice of a proper agglomerating liquid to provide for:
— strength of granules without impairing the granulation process;
— uniformly adjustable and stable spraying from injectors;
— retention of desired properties of source powder in granulated material.

The agglomerating liquid may be water, aqueous solutions of urea, vinyl acryl, anionic melamine-formaldehyde resin, impregnating urea-formaldehyde resin, polyethylene oxide of different concentration, as well as several original preparations.

The choice of particular low- and high-molecular compounds relating to different classes is due to their properties which make it possible to enhance production effectiveness, for example, the ability to combine with powdered aminoplast in pelletizing the material during the granulation process.

Owing to a wide range of properties and concentrations of the chosen solutions, the sprayed jet was suitably varied as to its width and the drop size, which is an apparent advantage in forming granulated material having a desired particle size.

To obtain moulding granulated material, use was made of a mist jet whereby small semone-like granules having a particle size of 1.0 to 1.2 mm were formed, while a jet having a drop size of 0.5 to 1.0 mm was used in production of casting granulated material

The formed granules were within 2.5–3.0 mm.

All the agglomerating solutions utilized (except for urea- and melamine-formaldehyde solution) provided for obtaining an adjustable-spray jet.

One of the criteria for evaluating quality of the obtained granulated material is the density of granules and their ability to withstand transportation.

In transit, the granules are subjected primarily to two types of load, more specifically, to vibration during shipment and to impact loads during handling operations.

The granule density may be evaluated considering the quantity of dust fraction (granules smaller than 0.18 mm) after the effect of a particular load.

Vibration loads may be simulated by the use of a vibrating chute at a vibration frequency corresponding to averaged conditions of transportation of the product by rail. Subjected to testing were samples of granulated material with different agglomerating liquid. The test samples in the amount of 0.5 kg (without dust fraction) were sealed in a polyethylene bag and loaded with a mass which (as referred to unit load) simulated actual conditions (freight car loading for transportation of a given material).

The test lasted for 5 and 12 hours. The quantity of dust fraction formed was recorded.

The test results are given in Table 8.

The analysis of the obtained data shows that the use of urea and vinyl acryl solutions fails to provide sufficiently strong granules (11 to 30% of the granules were disintegrated). For this reason, impact loads were applied to granules obtained by the use of water, polyethylene oxide, and several original preparations of agglomerating liquid.

The test procedure was as follows. Commercial granulated material in the amount of 25 kg was packed in sacks which were dropped from a 3-m height three times where-

Table 8. Effect of agglomerating liquid composition on strength of granulates in testing for vibration loads

Batch (agglomerating liquid used)	Amount of dust fraction (%) after testing during	
	5 hours	12 hours
1	2	3
Water	3.9	6.7
Urea aqueous solution:		
10%	15.6	30.0
20%	13.2	28.1
30%	14.0	27.4
Vinyl acryl aqueous solution:		
0.5%	11.2	14.0
1.0%	7.8	13.2
2.0%	6.9	11.1
Polyethylene oxide aqueous solution:		
0.006%	0.6	1.6
0.012%	0.6	2.8
0.018%	0.9	5.1
0.03%	1.2	5.4
0.1%	6.1	8.1
0.25%	5.2	7.2
0.5%	5.4	6.9
No. 3	0.4	1.7
No. 4	0.5	3.1
No. 9	0.4	1.6

Table 9. Effect of agglomerating liquid composition on granule strength in impact load tests

Batch (agglomerating liquid used)	Amount of dust fraction (%) aftertesting (number of impacts)	
	3	8
Water	5.4	10.8
Polyethylene oxide aqueous solution:		
0.006%	4.0	7.6
0.012%	3.0	5.7
0.018%	3.0	8.6
0.03%	2.9	4.2
0.1%	4.2	8.1
0.25%	4.2	7.6
No. 3	1.1	2.0
No. 4	1.4	3.0
No. 9	0.9	1.4

upon dust particles were separated and weighed. Next, each sack was dropped from the same height another five times, dust particles were separated and weighed.

The test results are given in Table 9.

The analysis of data given in Tables 8 and 9 above shows that the strongest granules are obtained with 0.006, 0.012 and 0.03% polyethylene oxide aqueous solutions and special agglomerating liquid (preparations Nos. 3, 4, and 9).

However, it should be borne in mind that, due to poor solubility of polyethylene oxide powder in water, certain difficulties are encountered in preparing solutions on its base.

3.4.1 Moisture Content of Granulated Material

Inasmuch as the driving force of the granulation process is determined by the presence of a liquid phase, variations of its content appreciably affect the granulation process.

It has been experimentally shown that, in granulating the Soviet-made КФ A type-2 powdered aminoplast (GOST 9389-88), the optimal moisture content in the product at the output of the granulator amounts to 23–29%. In this case, the output of commercial fraction measuring 0.2 to 1.0 mm is within 82–89% (depending on a particular preparation of agglomerating liquid), while fractions measuring 0.2 to 0.63 mm amount to 67–72%.

Figure 11 shows the output of commercial granulated material versus its moisture content with agglomerating liquid No. 3 utilized in granulation. Characteristic curves are also typical of the other preparations of agglomerating liquid.

An increase in moisture content over an optimal value results in formation of lumps in granulated material, heavy sticking of material on the plate surface and subsequent breakdown of the stuck layer, a disadvantage substantially impairing the granulation and drying processes and decreasing the granulator output.

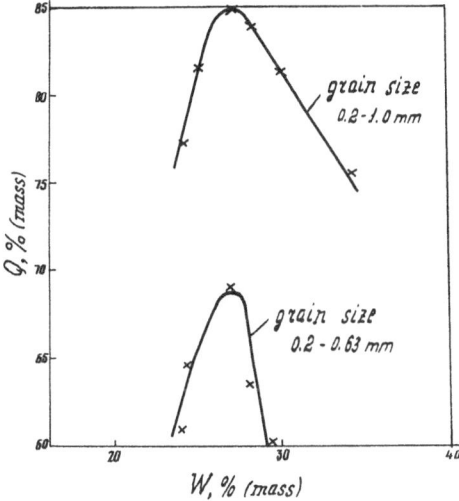

Fig. 11. Output of commercial granulated material Q versus its moisture content W

3.4.2 Angle of Inclination of Plate

The granule size depends, to a large measure, on the angle of inclination of the plate.

When the angle of inclination of the plate is within 53–57 %, the granulated material has a fraction of 0.2 to 0.63 mm. The granule size does not essentially change with the angle of inclination being varied within the above limits. Moreover, compaction of the granules is adequate. With a further increase in the angle of inclination, granules either fail to be formed or are too small, due to which the plate is heavily dusted with powdered aminoplast. Conversely, a decrease in the angle of inclination causes formation of larger granules and their greater compaction.

3.4.3 Rotational Speed of Plate

The size of the obtained granules is also affected by variations of the rotational speed of the plate with respect to a predetermined value (1.12 s^{-1}). Increasing the rotational speed of the plate to 1.39 s^{-1} causes the formation of smaller granules, a disadvantage attributable to the fact that the rolling speed required for obtaining a desired granule size is exceeded. In this case, the compaction process occurs at a faster rate.

Decreasing the rotational speed of the plate to 0.81 s^{-1} also results in formation of smaller granules.

3.5 Effect of Basic Drying Parameters of Granulated Aminoplast upon Its Technological Properties[1]

Drying is one of the most complicated stages in granulation of aminoplasts. As a matter of fact, drying entails a complicated physical and chemical process, in which removal

[1] The authors thank O. V. Vieletok and N. P. Beketova, who are research workers at NPO "Plastmassi" for their assistance in preparing this material.

of volatile constituents such as water, methanol and formaldehyde should occur in conditions precluding further condensation of CH_2OH functional groups (methylol products). An indication of the degree to which aminoplast is dried (one of quality indices) is presence of residual moisture, while an indication of condensation, which is also a quality index, is the presence of some methylol groups ensuring plastic properties of aminoplasts.

In the experiments use was made of a three-section fluidized-bed drier whose characteristics are given below.

To control the drying process and obtain quality granulated aminoplast, it is essential to know the relation between the final moisture content of the product and the temperature of the air coming out of the drier. Considering that aminoplast is a thermosetting material it is also of advantage to keep check on changes in the amount of methylol groups in the dried granulated material (as compared with the source material), depending on production parameters.

As the granulated aminoplast is dried, there occurs a polycondensation reaction, due to which the number of functional groups ($-CH_2OH$) changes. The drying conditions should, therefore, ensure production of material with the final moisture content not in excess of 3.5% by mass provided that the number of methylol groups therein remains constant or changes insignificantly.

In the course of drying, granulated material characterized by the initial moisture content of 25 to 32% by mass (after the granulator) and containing methylol groups within 9–12% is transferred to a uniformly fluidized state by hot air in the first zone of the drier. As the granulated material is dehydrated to 4–6% by mass, it enters the second zone. The moisture removed from the material in the first zone is surface (free) substance. The temperature of the material is constant and equal to the temperature of a wet thermometer, while the temperature of the outgoing air is close to the temperature of the material. The granulated material dried in the first zone contains up to 8–9% of methylol groups.

In the second zone of the drier diffusive moisture is removed from the material until residual moisture equals 1.5% by mass and the material is heated to a temperature of

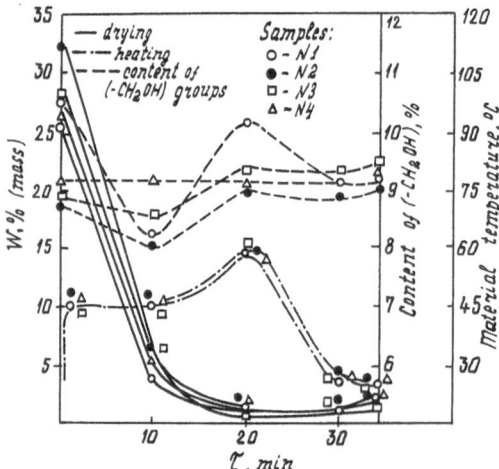

Fig. 12. Kinetics of drying and heating granulated aminoplasts with different agglomerating additives (samples 1 through 4) and changes in content of methylol groups

58 to 62 °C. The temperature of the exhaust air is 12 to 15 °C above the temperature of the material. In this case, the granulated aminoplast contains 9 to 10% of methylol groups.

After the third zone, the material cooled to 25 ± 2 °C is sized.

The material is found for about 10 min in each zone of the drier.

Figure 12 shows the results obtained in drying four typical batches of granulated aminoplast differing as to the presence of agglomerating additives (batches Nos. 1, 2 and 3, 4) and their concentration in the agglomerating liquid solution (batches Nos. 1, 2 and 3, 4).

The analysis of the obtained results shows that the drying process has two stages corresponding to a fixed-speed drying section and a falling-speed drying section. The drying rate equals 2.4 to 3.0%/min during the initial period. Critical moisture content of the material is in the range of 4 to 6%. The higher the initial moisture content of the material and the lower temperature of the drying agent at the drier input, the greater will be the critical moisture content. The granulated material in the second zone of the apparatus has a maximum temperature of 58 to 62 °C.

The cooling of granulated aminoplast with inadequately dried air cases the moisture content of the material to grow to 1.9–2.4% by mass. The content of functional groups is decreased throughout the period of fixed-rate drying of batches Nos. 1, 2, and 3 of the granulated material. This decrease is attributable to the continuing over-condensation reaction in the carbamide-formaldehyde binder of the aminoplast.

As the bound moisture is removed and the material is heated to a temperature of 58 to 62 °C in the falling-speed drying section, the content of methylol groups in batches Nos. 1, 2, and 3 slightly increases, which may be due to the effect of agglomerating additives.

The cooling of the experimental batches of the granulated material, except for batch No. 1, occurs with an essentially constant content of the indicated groups ($-CH_2OH$).

In batch No. 4 of the granulated material, the content of methylol groups remains essentially constant throughout the drying and cooling processes.

When the temperature of the air coming out of the second zone is below 68 °C, the moisture content of the second zone is below 68 °C, the moisture content of the material obtained in drying will exceed 3.5% by mass and the content of functional groups therein will be in excess of 9.2%. Such a material is formed into lumps in transit and storage. Moreover, it may bulge when processed into articles, a feature adversely affecting quality thereof.

On the other hand, it has been experimentally proved that, at the temperature of the air coming out of the second zone being in excess of 80 °C, the moisture content of the granulated material is below 1.2% by mass, while the content of methylol groups is less than 8.8%. With such characteristics, the material may not flow and, thus, cannot be processed into articles.

Hence, the temperature of the material in the second zone of the drier should be maintained within 58–62 °C with a view to obtaining quality granulated material having a residual moisture content of 1.5 to 3.5% (which corresponds to 8.8–9.2% content of methylol groups in the given material). The temperature of the air coming out of the second zone is within 68–75 °C.

The above data fully comply with the investigation results described in Ref. [33].

Table 10 gives experimental data averaged as a result of three measurements and

Table 10. Moisture content W_k of granulated material versus temperature T_2

Air temperature, T_2m °C	60	65	70	75	80	80	90
Final moisture content, W_k, %	6.0	4.0	2.8	1.75	1.2	0.85	0.55

Table 11. Content of methyl groups in granulated material versus temperature T_2

Air temperature T_2, °C	60	65	70	75	80	85
Content of CH_2OH, %	9.15	9.2	9.1	9.0	8.8	8.4

obtained in a fluidized-bed drier to illustrate variations of final moisture content W_k of granulated material depending on temperature T_2 of air coming out of the second drying zone. For the process in question, W_k is a single-valued function of T_2.

Entering the final moisture content permits determining the temperature of the outgoing drying agent and regulating in accordance with it the supply of moist granulated product by the use of the following equation:

$$T_2 = 81 \ exp \ (-0.051 \ W_k) \tag{1}$$

wherein T_2 is temperature of air coming out of the second drying zone, K; and W_k is final moisture content of granulated material, % by mass.

Table 11 gives experimental data relating to the content of methylol groups in granulated aminoplast versus temperature T_2.

Considering the fact that the content of methylol groups in the source powder is 9.1% it is apparent from the tabulated data that such a content of methylol products in dry granulated material may be obtained at the outgoing air temperature of 70 °C. Consequently, the final moisture content of the granulated material equal to 3.5% by mass may be provided when the temperature of the air coming out of the second drying zone is close to 70 °C. At this temperature, no change occurs in the content of functional groups and original plasticity of the material is preserved.

4 Granulation Process Lines

4.1 Granulation Techniques Involving Use of Plate Granulator

The technology developed and put into practical use in the USSR and relating to production of granules from powdered aminoplast by the use of a plate granulator comprises the following main stages:
— preparation agglomerating liquid;
— granulation;
— drying;
— sizing.
   The proposed process line is shown in Fig. 13.

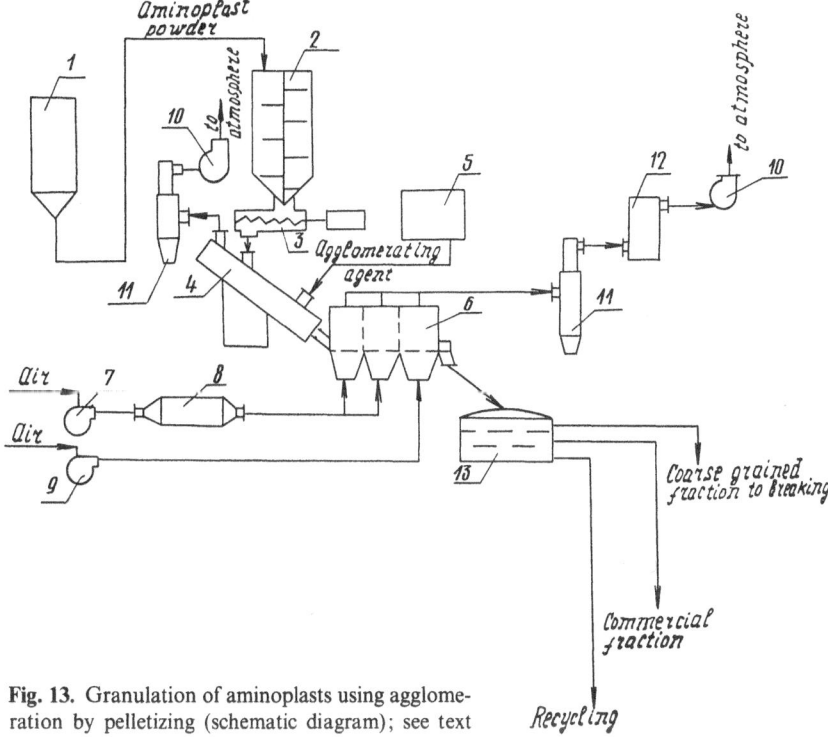

Fig. 13. Granulation of aminoplasts using agglomeration by pelletizing (schematic diagram); see text

4.1.1 Preparation of Agglomerating Liquid

The preparation of agglomerating liquid essentially consists in dissolving in water contained in service tank (5) the constituents facilitating granule formation and providing desired technological characteristics.

4.1.2 Granulation

Aminoplast is pneumatically transferred from bin (1) through a feeder and an injector to bin (2) provided with a cyclone separator and an agitator.

From bin (2) the aminoplast is supplied by screw feeder (3) to granulator (4).

The granulator comprises a moveable plate, a hermetically sealed enclosure having pipe connections to let the powder in and discharge dust-laden air, a plate drive, and an arrangement for spraying agglomerating liquid, which represents a bank of hydraulic injectors. The granulator structure permits changing the angle of inclination of the plate.

Agglomerating liquid is supplied from tank (5) by a proportioning pump through spray injectors to the granulator.

The quantity of liquid should amount to 25–30 % of the quantity of powder supplied. A desired liquid flow in each injector is set by the use of a flowmeter. The liquid should be uniformly sprayed from the injector outputs.

As particles of agglomerating liquid get into powdered aminoplast, the granulator forms round particles (granules) which are continuously removed from the granulator and flow by gravity via a sleeve to a sluice feeder and therefrom to drier (6), in which excessive moisture is removed from the product.

The process is controlled primarily by changing the amount of agglomerating liquid supplied.

When a visual inspection shows that the granules are too small, the liquid flow should be increased by 10 to 15 % for 3 to 5 min. If the granules are too large, the amount of the liquid supplied should be decreased by 10 to 15 %. Then one must readjust for the previousley set liquid flow as soon as the granule size is stabilized.

4.1.3 Drying of Granules

Drier (6) is designed for removing excessive moisture from granules and cooling them. It comprises air feeding and distributing assemblies (7) and (9), air heaters (8) and an assembly for removing dust from the air after drying.

The drier also includes instruments for measuring the temperature of material in a fluidized bed, the temperatures of incmoming and outgoing air in respective sections, a differential pressure. Also provided are inspection and cleaning holes and burst-proof membranes.

The granules coming to the drier are dried in a fluidized bed of hot air moving gradually from the input to the output along the bottom.

Before moist granulated material is fed to the drier, an underlayer of dry granulated material is normally provided in its sections. When the compound is supplied from the granulator to the drier, moist granulated material is left in the first zone for 10 min, after which the partition separating the first and second zones is removed and the material is transferred to the second zone. The mater remains there for another 10 min, after which the partition separating the second and third zones is opened and it enters the third zone. Following a 10-minute cooling period, the material is unloaded from the drier and fed by the sluice feeder to a sizer. Next, the material unloading process is continuously repeated.

The temperature of the material is 45 to 50 °C in the first zone and 58 to 62 °C in the second zone. The incoming air temperature is 85 to 100 °C in the first zone and 110 to 120 °C in the second zone.

In drying, the temperature at the inputs of the sections is maintained within predetermined limits by changing the temperature of the heat-transfer medium. It is necessary to ensure a stable supply of moist material to the drier and unloading of dry material therefrom. Also, the height of the material layer in the drier should be constant.

Aminoplast is cooled in the third section of the drier. Cooling air is taken in through a filter by fan (9).

To prevent environmental pollution, the air removed from the drier by fan (10) is cleaned in cyclone separator (11) and bag filters (12).

4.1.4 Sizing of Granules

Granulated material is continuously fed from the drier by the sluice feeder to the vibrating screen sizer (13).

The sizer has three branches adapted to hod large, small, and commercial fractions.

The commercial fraction of granulated aminoplast is supplied from the sizer to the bin and then through the feeder for packing into bags. The large and small fractions are collected in the wheeled bin and periodically delivered to ball mills to be crushed and subsequently agglomerated.

4.1.5 Main Assemblies of Granulation Line

The Soviet-made granulation lines comprise the following main assemblies (Fig. 13): screw feeder (3), granulator (4), fluidized-bed drier (6) and sizer (13).

The feeder comprises a cylindrical casing provided with loading and unloading branch pipes and accommodating a transfer screw supported in antifriction bearings. The screw is provided with a blade to loosen the material in the loading zone. The feeder drive is a variable-speed drive motor. The feeder output is adjusted manually with the help of the variable-speed drive by changing the angular speed of the transfer screw.

The specifications of the feeder are as follows: output — 0.61 to 3.65 m^3/h; screw diameter — 100 mm; pitch — 100 mm; screw speed — 3.04 to 18.4 s^{-1}.

The granulator comprises a rotary plate mounted on a shaft in a pivoted-frame bearing assembly. The plate is rotated by a drive roller which makes contact with its contact surface. The roller drive turns on the motor, V-belt transmission, and reduction gear.

The granulator has an enclosure to prevent ingress of dust in the production area.

The specifications of the granulator are as follows: output (as referred to souêce Mounted on the cover of the enclosure are a source powder feed connection, a ventilating connection, several injectors for feeding agglomerating liquid and an inspection hole cover. Provision is also made for a mechanism adapted to change the angle of inclination of the plate within 45–60°.

The granulator incorporates a novel arrangement [34] for removing stuck material from the working surfaces of the plate.

The vibration sizer classifying dry granulated aminoplast into three fractions is composed of three sections, two of which are provided with screens having different mesh powder) — 120 to 250 kg/h; plate diameter — 1800 mm; plate speeds — 0.81 s^{-1}, 1.12 s^{-1}, 1.39 s^{-1}; angle of inclination of the plate — 45 to 60°; flow of agglomerating liquid — 30 to 80 l/h.

The fluidized-bed drier is a welded square chamber accommodating meshes and separated by two vertical partitions into three sections having individual inputs suited to receive a heat-transfer agent. Two of the sections are used for drying granules and the third one is a cooling section. The granules are loaded and unloaded by sluice feeders.

The specifications of the fluidized-layer drier are as follows: output (as referred to moist product) — 300 ± 30 kg/h; output (as referred to evaporated liquid) — 50 ± 15 kg/h; capacity — 8 ± 0.2 m^3; drying air flow — 12.000 ± 500 m^3/h; cooling air flow — 3000 ± 200 m^3/h; mass — 3700 kg.

The vibration sizer classifying dry granulated aminoplast into three fractions is composed of three sections, two of which are provided with screens having different mesh sizes, while the third one has a bottom suited to remove the smallest fraction.

The sizer sections are joined together by yokes and the entire unit freely oscillates due to the action of a vibrator, the necessary support being provided by cylindrical springs. The vibrator shaft mounts two unbalanced masses whose resultant force causes vibrations in horizontal and vertical planes. The vibrator is operated from a motor through V-belt transmission.

The specifications of the sizer are as follows: output — 300 ± 30 kg/h; sieving area
— 1.4 m²; mesh size — 1.1 mm; 0.18 mm; mass of oscillating parts — 430 ± 5 kg.

4.2 Granulation Techniques Involving Use of Screw Mixer-Extruders

Standard production lines for obtaining granulated aminoplast by the use of screw
mixer-extruders are shown in Fig. 14a, b. The production line illustrated in Fig. 14a
is the Buss line for obtaining crushed material [35].

The powdered aminoplast is supplied from ball mill (1) to buffer tank (2) with an agitator and then to feed hopper (3) of mixer (4) wherein it is processed until a homogeneous
plastic extrudate is obtained.

This stage is to comply with stringent requirements for precise proportioning of thermal and pressure effects on processed material, which may be achieved by selecting
worm section profiles and temperature-speed parameters.

The material aggregate is fed from the mixer via a connecting shaft to the cooler (5)
and then to the crusher (6) wherein it is crushed preliminarily (by a single-roll crusher
unit) and finally (by a toothed-disc mill).

After crushing, the granulated material is fed to a sieve sizer (7), in which it is classified into fractions. Commercial fraction is supplied to bin (8), while other fractions are
returned to the ball mill (1) to be ground repeatedly. The granulated material is then
fed from bin (8) to the packing assembly (9).

A standard equipment set shown in Fig. 14b is used for manufacturing granulated
material [35].

From a raw-material loading assembly (or a grinding assembly) a mixture of appropriate constituents is fed at a constant speed to an intermediate tank (2) by a metering
unit. The raw material passes through a magnetic separator wherein metallic inclu-
sions are removed.

Tank (2) is provided with a vertical agitator, indicators of lower and upper levels of
material and an air-operated discharge valve.

The source material is supplied from tank (2) to the feed hopper (3) of mixer (4). The
feed hopper serving to continuously feed a homogenizing screw of mixer (4) comprises
a vertical agitator with blades in its upper portion to prevent sticking of the product
on the hopper walls. Arranged in the lower portion of the agitator is a separable screw

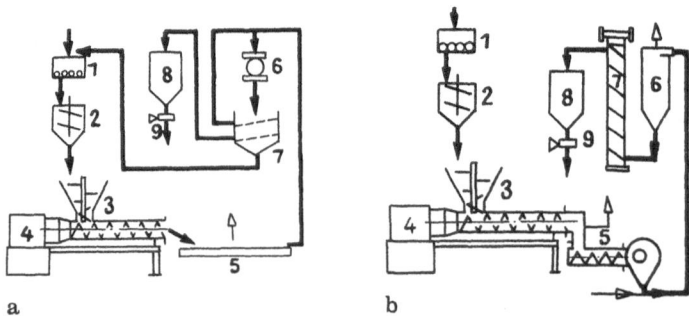

a   b

Fig. 14a, b. Granulation with Worm Mixer-Granulators for obtaining product in the form of:
(a) aggregate; (b) granules; see text

enabling continuous feed of the mixture. The agitator speed is adjusted within desirable limits. A built-in tachogenerator supplies speed data to a meter.

In the cylinder of the homogenizing screw the source product is transferred to a viscous-flow state and homogenization is accomplished owing to friction and additional heat provided by a heat-transfer medium. The temperatures in respective zones are as follows: 20 to 25 °C in zone I; 40 to 60 °C in zone II; 60 to 90 °C in zone III. To provide for heat supply, the line includes several heat-generating plants which allow creation of different temperature conditions in the screw zones and in the granulator nozzle body (within 90–100 °C). In the granulating screw the mixing occurs at 80 to 90 °C. The temperature of all the zones is automatically maintained at a constant level, the control being effected by means of electronic temperature controllers.

Initially, when the mixer (4) is started water may be fed to its cylinder by a proportioning pump, which is done to increase flow of the material. Water is supplied through a hole in a kneading tooth of the screw. The quantity of water is decreased as the equipment picks up steady-state operating conditions. The water contained in the mixer is then evaporated and may be removed from the connecting shaft by means of a vacuum pump.

The volatile substances formed in homogenization are drawn off by a vacuum pump through a filter in the vacuum treatment zone which is simultaneously used as the connecting shaft between the homogenizing and granulating screws.

The granulating screw forces out the obtained mixture through dies in a mesh. The formed strings are cut into granules on the mesh with a special knife operated from an independent drive. The rotational speed of the knife is continuously adjusted.

From the granulating screw (5) the granulated product is fed to an ejector, whence it is transferred by the air flow created by a fan to a cycline separator (6), in which the granulated material is separated from the dust-air mixture. The dust is passed to the filters and the granules are fed to a spiral oscillating conveyer (cooler) (7).

Oscillating motion of the conveyer causes reciprocating motion of the material which is unloated at the upper point of the conveyer. The granules are cooled and dried by clod air supplied by the fan. The cooling air temperature is controlled within 20–40 °C.

The cooled granulated material is supplied to bin (8) and then to the packing assembly (9).

The afore mentioned production lines comprise heat-generating plants and cabinets accommodating various test instruments and controls, which provide for: control of the mechanisms and devices in local and automatic blocked control modes; indication and digital recording of preset and current values of monitored parameters in the natural system of units, periodically and in response to a call; automatic warning and protective interlocking.

5 Properties of Granulated Aminoplasts

The above work package involving development and utilization of facilities and techniques for granulating aminoplasts by the use of a plate granulator has made it possible to set up production of granulated aminoplasts for commercial needs.

Tables 12 and 13 give basic technological and operational characteristics of moulding granulated aminoplast, grade and also characteristics of source powdered material.

Table 12. Technological characteristics of moulding granulated aminoplasts, Grade КФА 2-ПрГ

Batch (preparation of agglomerating liquid)	Duration of viscous-plastic state at 140 °C, s	Viscosity of melted compound, $\mu \cdot 10^{-9}$ Pa. s at 140 °C	Flow, mm	Bulk density g/m³	Looseness, s	Content of methylol groups, %	Granulometric composition, % Fractions, mm	
							0.1–1.0	> 1.0
1	2	3	4	5	6	7	8	9
Source powder								
No. 1	25	0.85	120	—	—	9.1	81.3	17.9
No. 2	20	0.98	150	0.56	5	9.1	80.0	19.8
No. 3	20	1.05	170	0.57	5	9.06	86.6	13.4
No. 4	20	0.87	130	0.59	5	9.37	90.55	9.45
No. 5[a]	20	0.95	125	0.55	5	9.19	80.6	12.4
No. 6[a]	40	1.35	100	0.55	4	10.2	88.1	8.88
No. 7[a]	40	1.40	170	0.56	4.5	9.1	79.3	11.4
No. 8[a]	40	1.30	170	0.62	4.0	9.03	82.6	12.7
	40	1.60	160	0.52	4.0	9.1		

[a] In batches Nos 5, 6, 7, and 8, the agglomeration liquid used was water and 0.006, 0.012, and 0.018 % polyethylene oxide aqueous solutions, respectively.

Table 13. Operational characteristics of moulding granulated aminoplast, Grade КФА2-ПрГ

Batch (preparation of agglomerating liquid)	Heat stability, °C	Specific surface resistance, ohms	Specific volume resistance, ohms · cm	Loss tangent of dielectric		Permittivity		Breakdown voltage, kV/mm
				10^6 Hz	50 Hz	10^6 Hz	50 Hz	
1	2	3	4	5	6	7	8	9
Powder	122	$(1.2-2.9) \times 10^{14}$	1.9×10^{14}	0.03	0.019	6.7	5.12	16.7
No. 1	115	1.8×10^{15}	5.7×10^{13}	0.031	0.023	7.04	6.35	16.1
No. 2	115	1.0×10^{15}	3.1×10^{13}	0.031	0.028	7.1	6.8	15.6
No. 3	116	1.7×10^{15}	4.5×10^{13}	0.033	0.023	6.5	6.6	16.5
No. 4	120	1.3×10^{15}	1.1×10^{14}	0.031	0.022	6.8	6.4	15.9
No. 5[a]	127	1.0×10^{16}	5.7×10^{14}	0.026	0.015	5.1	6.8	14.6
No. 6[a]	118	4.4×10^{15}	1.7×10^{14}	0.024	0.028	5.3	6.9	14.5
No. 7[a]	118	1.3×10^{15}	1.0×10^{14}	0.024	0.023	5.5	7.3	13.9
No. 8[a]	118	2.0×10^{14}	4.0×10^{13}	0.024	0.036	5.7	7.2	13.0

[a] In batches Nos 5, 6, 7, and 8, the agglomerating liquid used was water and 0.006, 0.012, and 0.018 % polyethylene oxide aqueous solutions, respectively

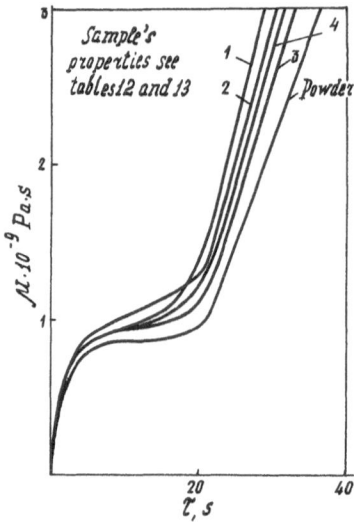

Fig. 15. Viscosity (μ) versus time (τ). Plastograms of powdered aminoplast and different samples of granulated materials on its base

The analysis of data give in Table 12 shows that viscosity of the material at 140 °C (moulding temperature) increases for all batches of granulated material, as compared with viscosity of source powder. This increase is apparently due to the use of different agglomerating liquids and specific constituents.

The duration of a viscous-plastic state at 140 °C is shorter in the case of batches Nos. 1 through 4, as compared with powder. This index is substantially higher for batches Nos. 5 through 8, as compared with powder.

Figure 15 shows plastometric curves characterizing moulding granulated material.

Granulated material is generally characterized by good technological indices, more specifically, by stable granulometric composition, adequate looseness and bulk density.

In addition to good physical and mechanical characteristics, moulding granulated material possesses adequate electrical characteristics (Table 13).

The analysis of data given in Tables 12 and 13 makes it possible to conclude that the composition of agglomerating liquid has a predominant effect on the properties of the obtained granulated material, which is confirmed by analyzing the characteristics of batches Nos. 1, 3, and 4 prepared by the use of preferred agglomerating liquids which the authors consider to be the most promising in the related field.

The Soviet-made moulding aminoplast КФА2-ПрГ corresponds to or even surpasses moulding aminoplasts UF, A11, MC and ISO 2112-77 commercially produced by the leading foreign firms, as regards basic operational properties.

6 Conclusion

The conducted investigations and also the analysis of the known methods made it possible to choose the most promising techniques and develop technical solutions enabling commercial production of granulated aminoplasts.

Such techniques include granulation by the use of screw mixer-granulators and granulation by pelletizing by the use of plate granulators.

Granulation with screw mixer-granulators requires fairly simple equipment. However, stringent requirements are placed on source material, as regards increased duration of a viscous-flow state at moulding temperatures and the need to rigidly maintain predetermined temperature conditions in all zones of the operating equipment.

The leading firms in developing and employing the aforementioned techniques are Buss, Werner & Pfleiderer, Berstorff, and some others.

The method described is not used on a wide scale in the USSR due to lack of basic production facilities meeting quality requirements.

Granulation by pelletizing with a plate granulator requires fairly simple equipment and permits large-scale production on one process line.

Requisite facilities and technology were developed and utilized in the USSR and abroad, for example, by Eirich company. The obtained granulated material possesses desired properties.

7 References

1. Kunststoffe (1986) 76: 1110
2. Plast Techn (1976) 22: 49
3. Fainshtein E. B. Aminoplastics (1985) Prospect VNIITEKhIM, Cherckassy, p 3
4. Zagryadskaya NF et al. (1976) Trends in thermosets granulation methodes Ser. "Plastics and Synthetic Resins" NIITEKhIM, Moscow, p 7
5. Mambish EJ et al. Granulation of thermosets (aminoplastics) (1979) Ser. "Plastics and Synthetic Resins" NIITEKhIM, Moscow, p 7
6. Mambish EJ et al. (1978) Manufacturing and Processing of Plastics and Synthetic Resins, NII-TEKhIM, Moscow, 1:13
7. Sockolow AD et al. (1970) USSR Inventors Certificate 268629
8. Gagarina EJ et al. (1973) USSR Inventors Certificate 379599
9. Kalinchev EL et al. (1982) USSR Inventors Certificate 914581
10. List X (1970) GB Pat 1215977
10. List X (1970) GB Pat 1215977
11. Hellstrom B et al. (1972) GB Pat 1279695
12. Abramov BJ et al. (1979) USSR Inventors Certificate 806435
13. Takewell R et al. (1977) US Pat 4050869
14. Mixers-Granulators Schugi (1978) Schugi Prospect
15. Maharinsky EG et al. (1967) USSR Inventors Certificate 208248
16. Furnet J (1968) USSR Pat 289547
17. Isonard Smith (1956) US Pat 2755509
18. Weihrauch E et al. (1973) FRG Pat 1769532
19. Weihrauch E et al. (1970) Swiss Pat 481734
20. Heinz H. et al. (1972) FRG Pat 2107927
21. Gresch W et al. (1967) Swiss Pat. 426215
21. (1967) Swiss Pat 426215
22. Ube Kosan KK (1972) Jap. Publ. 47-7442
23. Fridman ML (1986) Development and Improvement of equipment for mixing of thermoplastic melts, ZIINTIKhIMNEFTEMASH, Moscow, p 66
24. Paudal Co, Ltd Broschure, Japan (1974)
25. Berstorff Broschure, FRG (1987)
26. Baker Perkins Broschure, GB (1987)
27. Buss Broschure, Swiss (1981)
28. Werner and Pfleiderer Broschure, FRG (1987)
29. Eirich Broschure, FRG (1987)

30. Unit for granulation of dry pouring materials (1986) LenNIIKEMMASH Broschure, Dzerginsk
31. Aairmatic Broschure, FRG f1974)
32. Klassen PV, Grishaev IG (1982) Fundamentals of granulation methods, Khimiya, Moscow, p 272
33. Tunkel VI et al. (1987) Manufacturing and Processing of Plastics and Synthetic Resins, NII-TEKhIM, Moscow, 4: 17
34. Myasnova NN et al. (1983) USSR Inventors Certificate 1 071 304
35. Buss Broschure, Swiss (1978)

Editor: M. L. Fridman
Received December 21, 1988

Author Index Volumes 1–93

Subject Index